Springer Tracts in Modern Physics
Volume 119

Springer Tracts in Modern Physics

Volumes 90–106 are listed on the back inside cover

107 **Electronic Excitations in Condensed Rare Gases**
By N. Schwentner, E.-E. Koch, and J. Jortner 1985. VIII, 239 pages

108 **Particles and Detectors** Festschrift for Jack Steinberger
Edited by K. Kleinknecht and T. D. Lee 1986. 91 figs. X, 291 pages

109 **Metal Optics Near the Plasma Frequency**
By F. Forstmann and R. R. Gerhardts 1986. 30 figs. VIII, 132 pages

110* **Electrodynamics of the Semiconductor Band Edge**
By A. Stahl and I. Balslev 1987. 42 figs. IX, 215 pages

111 **Surface Plasmons on Smooth and Rough Surfaces and on Gratings**
By H. Raether 1988. 113 figs. X, 136 pages

112 **Tests of the Standard Theory of Electroweak Interactions**
By C. Kiesling 1988. 86 figs. X, 212 pages

113 **Radiative Transfer in Nontransparent Dispersed Media**
By H. Reiss 1988. 72 figs. X, 205 pages

114 **Electronic Transport in Hydrogenated Amorphous Semiconductors**
By H. Overhof and P. Thomas 1989. 65 figs. XIV, 174 pages

115 **Scattering of Thermal Energy Atoms from Disordered Surfaces**
By B. Poelsema and G. Comsa 1989. 74 figs. VIII, 108 pages

116 **Optical Solitons in Fibers**
By A. Hasegawa 1989. 22 figs. X, 75 pages

117 **Determination of Hydrogen in Materials** Nuclear Physics Methods
By P. K. Khabibullaev and B. G. Skorodumov 1989. 38 figs. X, 87 pages

118 **Mechanical Relaxation of Interstitials in Irradiated Metals**
By K.-H. Robrock 1990. 67 figs. VIII, 106 pages

119 **Rigorous Methods in Particle Physics**
Edited by S. Ciulli, F. Scheck, and W. Thirring
1990. 21 figs. X, 220 pages

120* **Nuclear Pion Photoproduction**
By A Nagl, V. Devanathan, and H. Überall
1991. 53 figs., Approx. 270 pages

121 **Current-Induced Nonequilibrium Phenomena in Quasi-One-Dimensional Superconductors**
By R. Tidecks 1990. 109 figs. IX, 341 pages

* denotes a volume which contains a Classified Index starting from Volume 36

S. Ciulli F. Scheck
W. Thirring (Eds.)

Rigorous Methods in Particle Physics

With 21 Figures

Springer-Verlag
Berlin Heidelberg GmbH

Professor Dr. Sorin Ciulli
Department of Mathematical Physics, Université de Montpellier II,
F-34095 Montpellier, France

Professor Dr. Florian Scheck
Institut für Physik, Johannes-Gutenberg-Universität Mainz, Staudingerweg 7,
D-6500 Mainz, Fed. Rep. of Germany

Professor Dr. Walter Thirring
Institut für Theoretische Physik der Universität Wien, Boltzmanngasse 5,
A-1090 Wien, Austria

Manuscripts for publication should be addressed to:

Gerhard Höhler
Institut für Theoretische Kernphysik der Universität Karlsruhe, Postfach 6980,
D-7500 Karlsruhe 1, Fed. Rep. of Germany

Proofs and all correspondence concerning papers in the process of publication should be addressed to:

Ernst A. Niekisch
Haubourdinstraße 6, D-5170 Jülich1, Fed. Rep. of Germany

ISBN 978-3-662-15073-3

Library of Congress Cataloging-in-Publication Data. Ciulli, S. (Sorin), 1933–. Rigorous methods in particle physics/S. Ciulli, F. Scheck, W. Thirring . p.cm. – (Springer tracts in modern physics; v. 119) Includes bibliographical references and index. 1. Particles (Nuclear
ISBN 978-3-662-15073-3 ISBN 978-3-540-47176-9 (eBook)
DOI 10.1007/978-3-540-47176-9
physics) – Mathematics. I. Thirring, Walter E., 1927–. II. Scheck, Florian, 1936–. III. Title. IV. Series. V. Series: Springer tracts in moderns physics; 119. QC1.S797 vol. 119 a [QC793.2] 530 s–dc20–[539.7'2] 90–10389

Originally published by Springer-Verlag Berlin Heidelberg New York in 1990
Softcover reprint of the hardcover 1st edition 1990

57/3140-543210 – Printed on acid-free paper

Dedication

André Martin has been a member of the theoretical group at CERN almost since its beginning and has been instrumental in setting the style of theoretical physics at CERN. On the one hand, he was never content with uncontrollable approximations and accepted only clean mathematics. On the other hand, he would not build abstract mathematical edifices which were not grounded on experimental facts. In fact, he lovingly collected experimental numbers to confront his theoretical results with. His charming personality made collaboration easy and so he ideally fulfilled the expectations one might have for people in such a distinguished position. Small wonder that many of his friends and collaborators wanted to dedicate a paper to his 60th birthday as a token of friendship and gratitude. It is with great pleasure that we introduce this beautiful collection of papers related to André Martin's work.

Preface

A consistent feature of André Martin's work has been the determination with which he has related his mathematical endeavours to real problems in physics. This volume of contributed papers, written on the occasion of André's sixtieth birthday, covers several important streams in mathematical physics as applied to modern elementary particle physics, reflecting the range of his scientific achievement.

Several articles deal with rigorous results in potential theory and its application to the spectrum of hadrons. Others update the use of analyticity in quantum field theory and particle physics.

Inverse problems and their ramifications form the subject of some articles dealing either with the general framework of the theory or with some specific questions relevant to particle physics.

Nonperturbative approaches to hadron physics, by way of variational improvements of perturbation theory and sum rules for quantum chromodynamics, are followed by topics on field theory addressing bifurcation theory, chiral models and solvable models.

The book concludes with a phenomenological analysis of Higgs particle interactions and a review of geometrical approaches to particle physics dealing with physics in higher dimensions as well as chiral anomalies in gauge theories.

This volume provides an overview of recent developments in mathematical physics which should be useful to theoretical physicists in general, not only those specializing in elementary particle physics. It will also be of interest to experimental particle physicists and to advanced research students.

August 1990

Sorin Ciulli
Florian Scheck
Walter Thirring

Contents

A Model for a Dia-Electric

H. Narnhofer and W. Thirring

Institut für Theoretische Physik der Universität Wien,
Boltzmanngasse 5, A-1090 Wien, Austria

André Martin is the master of the Schrödinger equation. Where others made do with cheap approximations he would not buy anything but a conclusive result proved by hard analysis. In recent years he applied his skill to meson spectroscopy where one deals with confining potentials. As a birthday present we will offer him a charged medium which gives such a potential purely in the framework of electrodynamics. We realize that this is just a toy and not the mechanism realized in nature. There it is supposed to emerge from QCD where also nonlinear and quantum effects are important. But our model may provide a formal basis to the heuristic discussions one finds in textbooks [1,2].

The Higgs field is a model for a perfect diamagnetic fluid. Its current obeys the relativistic London equations of superconductivity. They express that the current can follow freely the electric field and a magnetic field induces eddy currents which tend to cancel it. The coupled Maxwell-London equations are most concisely written in the notation of differential geometry [3] where the various fields are considered as elements of E_p, the p-forms. As derivations appear the exterior derivative $d : E_p \to E_{p+1}$ and its adjoint $\delta : E_p \to E_{p-1}$. Then the electromagnetic field $F \in E_2$, the Higgs current $J \in E_1$ and an external current $j \in E_1$ obey

$$\delta F = J + j, \qquad dJ = \mu^2 F, \qquad dF = \delta J = \delta j = 0 \tag{1}$$

(μ^{-1} is the penetration length). It follows $(d\delta - \mu^2)F = dj$ or if we work in $\mathbf{R}^4$ then by Fourier-transform ($k^2 = \vec{k}^2 - k_0^2$)

$$\tilde{F}(k) = -\frac{\widetilde{dj}}{k^2 + \mu^2} = -\frac{\widetilde{dj}}{k^2}\,\varepsilon(k)^{-1} \tag{2}$$

which shows the exponential screening of all fields generated by j. In fact, the dielectric constant is

$$\varepsilon(k) = 1 + \frac{\mu^2}{k^2}$$

and goes to ∞ for $k \to 0$.

For the magnetic susceptibility κ one finds

$$\kappa = \frac{-1}{1 + k^2/\mu^2}, \qquad \varepsilon(1+\kappa) = 1,$$

such that $\kappa \to -1$ for $k \to 0$. To get $\varepsilon < 1$ and $\kappa > 0$ we have to pervert the situation and change the sign of μ^2. Then the charges of the Higgs current run opposite to the usual way. Since this would produce a tachion we stabilize the situation by coupling J to another field $G \in E_2$ such that J and G alone behave normally. Correspondingly the field equations for the perverted Higgs model are

$$\delta F = J + j, \qquad dJ = \mu^2(G - F), \qquad \delta G = J, \qquad dF = dG = \delta J = \delta j = 0. \tag{3}$$

From these we conclude

$$\delta dJ = -\mu^2 j \Longrightarrow d\delta d\delta F = (-\mu^2 + d\delta)dj.$$

Turning to Fourier space we have

$$\tilde{F} = -(\frac{\mu^2}{k^4} + \frac{1}{k^2})\widetilde{dj}. \tag{4}$$

This corresponds to $\varepsilon(k) = k^2/(k^2 + \mu^2)$ and for $k \to 0$ we have $\varepsilon \to 0$ or a perfect dia-electric. Similarly $\kappa = \mu^2/k^2$ and (for $k_0 = 0$) $\kappa > 0$, the stuff is paramagnetic. The static Green function becomes

$$\int d^3k e^{ik\vec{x}}(\frac{\mu^2}{k^4} + \frac{1}{k^2}) = \frac{1}{4\pi}(\frac{1}{r} - \frac{\mu^2}{2}r) + \text{an infrared divergent constant.}$$

Thus we get just the potential popular among the quark confiners.

Remarks

1. When one constructs the Lagrangian and Hamiltonian for (3) one finds that J and G contribute negatively to the energy. On a simplified scalar version of this mechanism we have previously shown [4] that one can quantize nevertheless with a positive metric in Hilbert space. However, there are no space-time translation invariant states and the time evolution is not unitarily implemented. Indefinite metric does not really help [5]. Nevertheless a unitary S-matrix might exist.

2. There are other classical models for a dia-electric. J. Hôsek has constructed one with charged gravitons [6].

3. This electrodynamic type of theory would be catastrophic for strong interactions since it would lead to van der Waals type potentials $\sim 1/r^4$ between two quark-antiquark pairs. How these long range forces are avoided in QCD has yet to be demonstrated.

References

1 T.D. Lee, Particle Fields and Introduction to Field Theory, Harwood Academic Publishers, Chur, London, New York (1981).

2 K. Gottfried, V.F. Weisskopf, Concepts of Particle Physics II, Oxford University Press,New York (1986).

3 W. Thirring, A Course in Mathematical Physics II, 2. ed., Springer, New York (1986).

4 H. Narnhofer, W. Thirring, Phys. Lett. 76B, 428 (1978).

5 U. Maschello, SISSA preprint 1989.

6 J. Hôsek, Talk at Triangle Meeting, Vienna (1989).

Potential Models for Quarkonium Systems*

H. Grosse

Institut für Theoretische Physik der Universität Wien,
Boltzmanngasse 5, A-1090 Wien, Austria

Abstract

We review some of the work done together with André Martin during the last twelve years. We concentrate especially on the problem of order of levels in potential models and indicate results on the ℓ-dependence of energy levels, on the wave function at the origin and on the inverse problem for confining potentials.

1 Introduction

Almost all questions, which we discussed during so many years were connected to one equation, which we may call André 's equation, but which is also called the radial Schrödinger equation on $L^2(\mathbf{R}^+, dr)$:

$$\left\{-\frac{1}{2m}\frac{d^2}{dr^2} + \frac{\ell(\ell+1)}{2mr^2} + \lambda V(r)\right\} u(r) = Eu(r). \qquad (1.1)$$

Although this equation is established since so many years new results have been obtained recently. We may label the energy levels by the angular momentum ℓ and the number of zeros of the wave function within $(0,\infty)$ n, and study the dependence of $\varepsilon_{n,\ell}$ as a function of m and λ. The quasiclassical limit corresponds to the strong coupling limit $\lambda \to \infty$. In that limit we expect that the total number of negative energy levels N_d becomes the classical phase space volume

$$N_d \overset{\lambda\to\infty}{\simeq} \int \frac{d^dx d^dp}{(2\pi)^d} |p^2 + \lambda V(x)|_- = Cl_d \int d^dx |\lambda V|_-^{d/2}, \qquad (1.2)$$

* *Dear André,*

I have had the great pleasure to work together with you on Schrödinger potential problems. It is an equal pleasure for me to review our common work on the occasion of your sixtieth birthday. I wish you many happy recurrences and I am convinced that you shall find many nice new proofs and results in your fields.

We celebrated your birthday on the 23rd of September in Annecy. We were very close to your birthday; indeed we obtained only an upper bound, but a quite good one. For me, you are the master of obtaining bounds. You ask, motivated by physics, very interesting questions, obtain often the way to prove such conjectures or cook, on the other hand, in a clever manner counterexamples.

where $|V|_-$ denotes the absolute value of the attractive part of V. (1.2) has been proven first for $d = 1$ by Chadan (1968), the result for arbitrary d goes back to Martin (1972).

During my first CERN visit a number of physicists, motivated by the stability of matter problem, tried to find bounds of the form

$$N_d \leq C_d \int d^d x |\lambda V|_-^{d/2}, \tag{1.3}$$

where Cl_d has been replaced by constants C_d. A new way of counting the number of bound states of a potential, invented by André, was the starting point of our investigation of this problem. As a function of ℓ, $\varepsilon_{n,\ell}$ is monotonically increasing and comes to zero energy at special values ℓ_k. If we denote by ν_ℓ the number of bound states of angular momentum ℓ we obtain following the Regge trajectories

$$\nu_\ell = \sum_{k=0}^{\nu_0 - 1} \Theta(\ell_k - \ell). \tag{1.4}$$

For three dimensions, for example, we may therefore express N_3 in terms of ℓ_k

$$N_3 = \sum_{k=0}^{\nu_0-1} \sum_{\ell=0}^{[\ell_k]} (2\ell + 1) = \sum_{k=0}^{\nu_0-1} ([\ell_k] + 1)^2, \tag{1.5}$$

and the r.h.s. of (1.5) can be bounded by an integral over $|V|_-^{3/2}$. See, for example, Grosse, Glaser and Martin (1978).

The main part of our collaboration was connected to quarkonium physics. After the discovery of the J/ψ and ψ' resonances and their interpretation as bound states of a charmed quark-antiquark system it became clear that a description in terms of nonrelativistic potential models seems to be appropriate. The standard potential $-\alpha_s/r + ar$ is motivated by QCD and confinement ideas. One finds a P-state between the first two S-states (J/ψ and ψ'), and a D-state above the ψ'. The question, first asked by Bég, was, how general is the observed order of levels in charmonium. This led to an investigation of

2 The Order of Levels

If we compare the spectra of the Coulomb problem and of the harmonic oscillator in three dimensions we expect for a convex combination of both potentials a number of relations for the spectral values. A multiplet of levels, which are degenerate for the Coulomb case becomes ordered such that $\varepsilon_{n,\ell} > \varepsilon_{n-1,\ell+1}$, the higher angular momentum state gets lower energy. We expect even more: If we introduce the main quantum number N and introduce $E_{N,\ell} = \varepsilon_{n,\ell}$ with $N = n + \ell + 1$ we expect that

$$E_{1,S} \leq E_{2,P} \overset{A}{\leq} E_{2,S} \overset{B}{\leq} E_{3,D} \overset{A}{\leq} E_{3,P} \leq E_{3,S}. \tag{2.1}$$

Some of the above inequalities are trivial. The centrifugal term shifts all levels up and therefore $E_{N,S} \le E_{N,P} \le E_{N,D} \ldots$ André's first result concerns inequalities A. It states that

$$\text{A)} \qquad \frac{d^3}{dr^3}(r^2 V(r)) \gtrless 0 \Rightarrow E_{2,P} \lessgtr E_{2,S}, E_{3,D} \lessgtr E_{3,P}. \tag{2.2}$$

For the proof one starts from the Coulomb potential, uses the Virial theorem and the Feynman-Hellmann theorem, and in addition the nodal structure of the involved wave functions enters. This makes it difficult to generalize the result to higher energy states. Since it is possible to map the Coulomb problem to the harmonic oscillator a similar result should hold for the latter case. Following the above ideas I realized that

$$\text{B)} \qquad \frac{d^3}{d\rho^3}\left(r^2 V(r)\right)\Big|_{\rho=r^2} \gtrless 0 \Rightarrow E_{2,S} \lessgtr E_{3,D}. \tag{2.3}$$

Examples of potentials, which give the order of levels as it is observed in the spectra of charmonium and bottonium are given, for example, by superpositions of the form

$$V(r) = \int_{-1}^{2} d\alpha r^{\alpha} \varepsilon(\alpha) \rho(\alpha), \qquad \rho \ge 0. \tag{2.4}$$

For a review of the early results see Grosse and Martin (1980).

Feldman, Fulton and Devoto (1979) studied the problem of level ordering within the WKB approximation. As a result, they found out that the Laplacian of the potential controls the way multiplets are splitted:

$$\Delta V \gtrless 0 \Rightarrow \frac{\partial E}{\partial \ell}\Big|_N \lessgtr 0. \tag{2.5}$$

The next step concerns the perturbation expansion around the degenerate cases. To a certain surprise it turns out that the Laplacian of the potential enters in an essential way (Grosse and Martin 1984):

Theorem: Let $H = -\Delta - \frac{\alpha}{r} + \lambda V(r)$ where $\Delta V(r) \gtrless 0$ for all $r \neq 0$ and denote by $\varepsilon_{n,\ell}(\lambda)$ the levels of H. Then

$$\delta_{n,\ell} := \frac{d}{d\lambda}(\varepsilon_{n,\ell}(\lambda) - \varepsilon_{n-1,\ell+1}(\lambda))|_{\lambda=0} \gtrless 0 \quad \forall n, \ell. \tag{2.6}$$

The main ideas of the proof are simple. We intend to determine the sign of the difference of level shifts given by

$$\delta_{n,\ell} = \int_0^\infty dr V(r)\{u_{n,\ell}^2(r) - u_{n-1,\ell+1}^2(r)\}, \tag{2.7}$$

where $u_{n,\ell}(r)$ denote the Coulomb wave functions. We observe furthermore that there exist ladder operators a_ℓ and $a_\ell^\dagger$ which map from $u_{n,\ell}$ to $u_{n-1,\ell+1}$ and vice versa. This is easily realized if we first factorize the radial Hamiltonian as

$$h_\ell = -\frac{d^2}{dr^2} + \frac{\ell(\ell+1)}{r^2} - \frac{1}{r} = a_\ell^\dagger a_\ell - \frac{1}{4(\ell+1)^2} \tag{2.8}$$

where $a_\ell^\dagger$ denotes the adjoint of a_ℓ and is given by

$$a_\ell = \frac{d}{dr} - \frac{\ell+1}{r} + \frac{1}{2(\ell+1)}. \tag{2.9}$$

The factorization was known already to Schrödinger and is connected to the O_4 symmetry. The projection of the Runge-Lenz vector onto states with magnetic quantum number $m = \ell$ becomes the differential operator (2.9). The commutator of operators a_ℓ, $a_\ell^\dagger$ and h_ℓ is easily calculated and becomes

$$[a_\ell^\dagger, a_\ell] = -2\frac{\ell+1}{r^2}, \qquad [h_\ell, a_\ell] = -2\frac{\ell+1}{r^2}a_\ell. \tag{2.10}$$

From the second relation we deduce the intertwining property of a_ℓ and $a_\ell^\dagger$ since $h_{\ell+1}a_\ell = a_\ell h_\ell$ and $h_\ell a_\ell^\dagger = a_\ell^\dagger h_{\ell+1}$. These mappings between the wave functions involved in (2.7) allow to rewrite $\delta_{n,\ell}$ in such a manner that the Theorem results.

Similar ideas, but perturbing around the oscillator yield the result for $H = -\Delta + r^2 + \lambda V(r)$ that

$$\frac{d^2}{d\rho^2} V(r)\big|_{\rho=r^2} \gtrless 0 \Rightarrow \frac{d}{d\lambda}(\varepsilon_{n,\ell}(\lambda) - \varepsilon_{n-1,\ell+2}(\lambda))\big|_{\lambda=0} \gtrless 0. \tag{2.11}$$

After a seminar in Vienna on that subject, a colleague became interested in the search for a nonperturbative result. After various attempts we found a very general result (Baumgartner, Grosse and Martin 1984) for the spectra of Schrödinger operators with radial symmetric potentials.

Theorem: Let $H = -\Delta + V(r)$ and assume that $\Delta V(r) \gtrless 0 \; \forall \, r \neq 0$. Then the spectrum is ordered so that $\varepsilon_{n,\ell} \gtrless \varepsilon_{n-1,\ell+1}$.

Three steps lead to this result:

a) We use the factorization property of second order differential operators, which is related to supersymmetric quantum mechanics. The generalization of eqs. (2.8) to (2.10) to the case of an arbitrary potential starts from the ground state wave function $u(r)$, which has no zeros within $(0, \infty)$. We define $A = d/dr + g(r)$, where $g(r) = -u'(r)/u(r)$ denotes the negative logarithmic derivative of $u(r)$. We obtain

$$H - \varepsilon := A^\dagger A = -\frac{d^2}{dr^2} + g^2(r) - g'(r) = -\frac{d^2}{dr^2} + V - \varepsilon, \tag{2.12}$$

where we used the Riccati equation obeyed by $g(r)$. (2.10) is replaced by the commutation rules

$$[A^\dagger, A] = -2g', \qquad [H, A] = -2g'A. \tag{2.13}$$

We deduce again an intertwining property since $(H+2g')A = AH$. This shows that the operators H and $H + 2g'$ are "essentially" isospectral. Their spectra coincide except for the ground state of H, which has been projected away. That $AA^\dagger$ and $A^\dagger A$ are "essentially" isospectral can be obtained easily by multiplying the eigenvalue equations by $A^\dagger$ and A, respectively. The above result goes back to a theorem of Darboux from 1882. We have clearly not been aware of the history of this subject. We have been like Monsieur Jourdain, "qui a fait la prose, sans savoir". Supersymmetric quantum mechanics starts by introducing supercharges Q and $Q^\dagger$, whose anticommutator yields the Hamiltonian:

$$Q = \begin{pmatrix} 0 & A \\ 0 & 0 \end{pmatrix}, \qquad Q^\dagger = \begin{pmatrix} 0 & 0 \\ A^\dagger & 0 \end{pmatrix}, \qquad \{Q, Q^\dagger\} = \begin{pmatrix} AA^\dagger & 0 \\ 0 & A^\dagger A \end{pmatrix}. \tag{2.14}$$

For more details see, for example, Grosse (1989).

b) The next step is a master piece of André Martin. We start from angular momentum ℓ, change g into $g_\ell = -u_\ell'(r)/u_\ell(r)$ and try to compare h_ℓ and $h_{\ell+1}$, where

$$h_\ell = -\frac{d^2}{dr^2} + \frac{\ell(\ell+1)}{r^2} + V(r). \tag{2.15}$$

From (2.13) we obtain "essential" isospectrality of h_ℓ and $h_\ell + 2g_\ell'$. How to compare $h_\ell + 2g_\ell'$ and $h_{\ell+1}$? Within a few days André told us the

Lemma: If $\Delta V(r) \gtrless 0 \ \forall\, r \neq 0$, then

$$g_\ell' \gtrless \frac{\ell+1}{r^2}. \tag{2.16}$$

Although the proof uses only "elementary" techniques like several times comparison potentials and Wronskian relations, it is rather tricky. We shall not give details here and refer to Baumgartner, Grosse and Martin (1984).

c) We combine (2.14) and (2.16) and obtain the operator relation

$$h_\ell + 2g_\ell' \gtrless h_{\ell+1} \Rightarrow \varepsilon_{n,\ell} \gtrless \varepsilon_{n-1,\ell+1}, \tag{2.17}$$

if the Laplacian of V has a definite sign for all $r \neq 0$.

Remarks: From the proof of (2.16) it follows that it is sufficient to assume $\Delta V < 0$ for some $r < R$ and $dV/dr < 0$ for $r > R$ in order to conclude that $\varepsilon_{n,\ell} < \varepsilon_{n-1,\ell+1}$.

The change of variables $r \to \rho = r^2$ and $u \to w = r^{1/2}u$ yields a new angular momentum $\lambda = (2\ell - 1)/4$ and connects Coulomb to oscillator. If we change even $r \to r^\alpha$ we obtain the

Corollary: Let $H = -\Delta + V$, $Hu_{n,\ell} = \varepsilon_{n,\ell}u_{n,\ell}$. If

$$\begin{aligned} D_\alpha V \geq 0 \quad &\text{for } 1 < \alpha < 2 &&\Rightarrow \varepsilon_{n,\ell} \geq \varepsilon_{n-1,\ell+\alpha} \\ D_\alpha V \leq 0 \quad &\text{for } \alpha > 2 \text{ or } \alpha < 1 &&\Rightarrow \varepsilon_{n,\ell} \leq \varepsilon_{n-1,\ell+\alpha} \end{aligned} \tag{2.18}$$

$$D_\alpha = \frac{d^2}{dr^2} + (5 - 3\alpha)\frac{1}{r}\frac{d}{dr} + 2(1-\alpha)(2-\alpha)\frac{1}{r^2}.$$

This result has been improved by A. Martin for power potentials.

It would be surprising if there would not exist a similar result for the continuous spectrum. Phase shifts for different angular momenta can be related to each other if ΔV has a definite sign.

It is interesting to note that the condition on the potential which implies a particular level ordering $\Delta V \gtrless 0$ gives a definite perihelion shift for the classical orbits too.

In the relativistic Dirac Coulomb problem levels to values $k = j + 1/2$ and $-k$ are degenerate. It is surprising that the splitting within the perturbation expansion is again controlled by the Laplacian of V.

Applications: We have found a number of applications for the result of the level ordering. For quarkonium physics we like to have $\Delta V > 0 \;\forall\; r \neq 0$. We write the force as $-V'(r) = -Z(r)/r^2$ and introduce an effective charge $Z(r)$. This r-dependent charge should therefore decrease for $r \to 0$ which is in agreement with asymptotic freedom ideas. For large r, $V(r)$ should grow like r and yields $\Delta V > 0$ too. In addition we like to get $\varepsilon_{n,\ell} < \varepsilon_{n-1,\ell+2}$ which is implied by $\frac{1}{dr}\frac{1}{r}\frac{d}{dr}V < 0$. The potential should therefore be increasing and concave, conditions which are implied by reflection positivity within lattice QCD.

During our studies we realized that the opposite sign of the Laplacian yields applications in atomic physics. Consider the Hartree equation for atoms with closed shells except one outer electron, then

$$(-\Delta - \frac{Z}{r} + \int \frac{d^3y\rho(\vec{y})}{|\vec{x} - \vec{y}|})\psi = E\psi \tag{2.19}$$

determines the spectrum, where ρ should be determined self-consistently. (2.19) is difficult to solve, but $\Delta V < 0 \;\forall\; r \neq 0$, which implies that $\varepsilon_{n,\ell} < \varepsilon_{n-1,\ell+1}$. If we even look to the periodic table and check how nature fills the shells of atoms we realize that after the $1S$ shell the $2S$ shell is filled before the $2P$ shell and after the $3S$ shell the $3P$ shell is filled in accordance to our level ordering. That the $3D$ state even overcomes the $4S$

state is responsible for many properties of materials but is clearly beyond the applicability of our simple results.

We mention that the spectra of mesonic atoms are in accordance with the above theorem too.

3 Further Results

It is clearly impossible to review the whole work on potential models done by A. Martin within the last 12 years. We shall mention just a few results.

One may consider the ℓ-dependence of energy levels. That $\varepsilon_{0,\ell}$ is concave as a function of $\ell(\ell+1)$ follows from the Min-Max principle. Concavity of $\varepsilon_{0,\ell}$ as a function of ℓ has been studied by Common and Martin who have shown that

$$\frac{d^2}{d\rho^2} V(r)\big|_{\rho=r^2} \gtrless 0 \Rightarrow \frac{d^2}{d\ell^2}\varepsilon_{0,\ell} \gtrless 0. \tag{3.1}$$

It is known that $\varepsilon_{n,\ell}$ need not be concave in ℓ for $n \geq 1$. The proof uses the Lemma, equ. (2.16), and a moment inequality. Combining with the level ordering theorem leads to conditions such that $\varepsilon_{1,\ell}+\varepsilon_{0,\ell} \gtrless 2\varepsilon_{0,\ell+1}$. Application to charmonium asserts that χ_c lies above the average of J/ψ and ψ'. The more general inequality $\varepsilon_{n+1,\ell} + \varepsilon_{n,\ell} \lessgtr 2\varepsilon_{n,\ell+1}$ for $n \geq 1$ is known to hold for perturbation expansions around r^2, in the limit $n \to \infty$ ℓ fixed and for n fixed and $\ell \to \infty$ under appropriate assumptions of the potential.

A further result on the ℓ-dependence follows from the Feynman-Hellmann theorem and the nodal structure of the involved wave function:

Theorem: If

$$\left(\frac{d}{dr^2}\right)^2 r^2 \frac{dV}{dr} \leq 0 \Rightarrow \frac{d}{d\ell}(\varepsilon_{1,\ell} - \varepsilon_{0,\ell}) < 0. \tag{3.2}$$

Another type of relations follows from the observation that increasing mass lowers all levels. We may therefore try to compensate this decrease by increasing the angular momentum. Use of the elementary inequality that $4\int_0^\infty u'^2 dr \geq \int_0^\infty dr u^2/r^2$ and the Min-Max principle led to the result that

$$\varepsilon_{n,\tilde{\ell}}(M) \leq \varepsilon_{n,0}(m) \text{ for } \tilde{\ell} = \frac{1}{2}\left(\sqrt{\frac{M}{m}} - 1\right), \tag{3.3}$$

where we indicated explicitly the mass dependence of the energy levels $\varepsilon_{n,\ell}(m)$. If the potential fulfills condition A of equ. (2.2) inequality (3.3) can be improved for $n = 0$ to

Theorem: If V fulfills condition A, it follows that

$$\varepsilon_{0,(\sqrt{M/m}-1)}(M) \leq \varepsilon_{0,0}(m). \tag{3.4}$$

Since the nodal structure of the ground state wave function together with a continuity argument enters, a possible generalization to higher levels is not obvious. For applications we assume flavour independence of the potential and may compare the charmonium to the bottonium ground state. This allows us to bind the mass difference $M_b - m_c \geq 3.29$ GeV.

In addition to energy levels, transition amplitudes and decay widths are measurable. The latter is proportional to the square of the wave function at the origin, which becomes the square of the reduced wave function at $r = 0$. Experimentally $\Gamma_{\psi \to e^+e^-} \simeq 4.8$ keV while $\Gamma_{\psi' \to e^+e^-} \simeq 2.1$ keV. A result of André Martin states that

$$\frac{d^2}{dr^2}V(r) \gtrless 0 \Rightarrow u'_{1,0}(0)^2 \gtrless u'_{0,0}(0)^2. \tag{3.5}$$

(3.5) is obtained from a continuity argument and using the node structure and the relation $u'^2(0) = \int_0^\infty dr u^2(r) dV/dr$. From the data an overall concave potential is preferred against an overall convex one.

Generalization of (3.5) to higher states is more difficult to prove: it holds within the WKB approximation, within first order perturbation theory around the linear potential and it is true asymptotically:

$$\text{if } \frac{d^2}{dr^2}V(r) \gtrless 0, \qquad \frac{dV}{dr} > 0 \Rightarrow u'_n(0) \overset{n\to\infty}{\longrightarrow} \left\{ \begin{array}{c} \infty \\ 0 \end{array} \right. . \tag{3.6}$$

The mass dependence of $u_{0,\ell} = u_\ell(m)$ is obviously of interest:

$$\frac{d^2}{dr^2}V(r) \leq 0 \Rightarrow \lim_{r\to 0}\left|\frac{u_\ell(m_1)}{u_\ell(m_2)}\right|^2 > \frac{m_1}{m_2} \quad \forall \ell. \tag{3.7}$$

Applying such a bound favours charge $|Q_b| = 1/3$ against $2/3$ for the bottom quark. From a possible generalization only partial answers are known.

We might ask about the mass dependence of the probability for a particle to be present within a sphere of radius R. For the ground state André found (after a seminar of B. Simon) a simple proof of a monotonicity result:

$$\text{if } \frac{dV}{dr} > 0 \Rightarrow \frac{\partial}{\partial m}\int_0^R dr |u_{0,0}(r)|^2 > 0. \tag{3.8}$$

Although the inverse problem has a long history it turned out that the problem for confining potentials was not treated before. For a fixed angular momentum especially $\ell = 0$, let $u_{n,0} = u_n$, we proved in Grosse and Martin (1979):

Theorem: Assume that there are given two sequences E_n and $u'_n(0)$ belonging to a confining potential which fulfills

$$V \in L^1_{loc}, \qquad \int_R^\infty dr \frac{V'^2(r)}{V^{5/2}(r)} < \infty, \qquad \int_R^\infty dr \frac{V''(r)}{V^{3/2}(r)} < \infty \text{ for some } R, \tag{3.9}$$

then $V(r)$ is uniquely determined.

We have related this problem to a simpler one, namely to the inverse problem for the same equation but with different boundary conditions

$$-w_n'' + (V(r) - \varepsilon_n)w_n = 0, \qquad w_n'(0) = 0. \tag{3.10}$$

The usual Gelfand-Levitan procedure is applicable to that problem. The relationship between both problems has been found by studying the quantity

$$R(E) = \left(\frac{d}{dr}u_E(r)/u_E(r)\right)\Big|_{r=0}, \tag{3.11}$$

where $u_E(r)$ denotes the solution of the Schrödinger equation defined by the WKB decay at infinity. $R(E)$ turns out to be a Herglotz function and admits a representation of the form

$$R(E) = C + E\sum_{n=0}^{\infty}\frac{|u_n'(0)|^2}{E_n(E_n - E)} \tag{3.12}$$

where one subtraction C is needed. The subtraction constant was determined by

$$C = \lim_{N\to\infty} C_N, \qquad C_N = \sum_{j=0}^{N}\frac{|u_j'(0)|^2}{E_j} - \frac{2}{\pi}\sqrt{\frac{E_N + E_{N+1}}{2}}. \tag{3.13}$$

The data determine $R(E)$ but then

$$-R^{-1}(E) = \sum_{n=0}^{\infty}\frac{|w_n(0)|^2}{\varepsilon_n - E} \tag{3.14}$$

yields the information necessary for the second problem.

We have treated also the inverse problem if the ground state energy is given as a function of ℓ.

It remains to be said that André has developed a number of nice results recently especially on energy differences, on bounds on moments of ground state density (together with A. Common), on bounds on the kinetic energy as well as on bounds on the three body binding energy (together with J.L. Basdevant and J.M. Richard).

Hopefully the toponium spectrum will allow further applications of the potential approach.

A more extensive treatment of all the above problems will be given in our forthcoming book (Grosse and Martin 1991).

References

Baumgartner, B., Grosse, H., Martin, A. (1984): Phys. Lett. **146B**, 363

Chadan, K. (1968): Nuovo Cimento **58A**, 191

Feldman, G., Fulton, T., Devoto, A. (1979): Nucl. Phys. **B154**, 441

Grosse, H. (1989): Supersymmetric Quantum Mechanics, Proc. of the Brasov Int. School 1989 on "Recent Developments in Quantum Mechanics".

Grosse, H., Glaser, V., Martin A. (1978): Commun. Math. Phys. **59**, 197

Grosse, H., Martin A. (1979): Nucl. Phys. **B148**, 413

Grosse, H., Martin A. (1980): Phys. Rep. bf60, 341

Grosse, H., Martin A. (1984): Phys. Lett. **134B**, 368

Grosse, H. Martin A. (1991): *Bound States in Schrödinger Potential Theory* (Cambridge Univ. Press, Cambridge 1991)

Martin, A. (1972): Helv. Phys. Acta **45**, 140

Some New Bounds on the Number of Bound States

K. Chadan[1] *and R. Kobayashi*[2]

[1]Laboratoire de Physique Théorique et Hautes Energies*, Université de Paris XI, F-91405 Orsay Cedex, France
[2]Department of Mathematics, Science University of Tokyo, Noda, Chiba 278, Japan

Abstract. In this work, we generalize the well-known bounds of Glaser, Grosse, Martin, and Thirring on the number of bound states of a particle in a spherically symmetric potential. The new bounds fill the gap between the above bounds, and the Calogero bound.

For a spherically symmetric potential V(r), it has been shown by Glaser, Grosse, Martin, and Thirring [1], that the number of bound states in the angular momentum state "ℓ" has the upper bound

$$n_\ell \leqslant C_{\ell,p} \int_0^\infty \left|r^2 V_-\right|^p \frac{dr}{r} \tag{1}$$

where the free parameter p has the range

$$1 \leqslant p \leqslant \frac{3}{2} \tag{2}$$

and

$$C_{\ell,p} = \frac{(2\ell+1)^{1-2p}(p-1)^{p-1}\,\Gamma^2(2p)}{p^p\,\Gamma^2(p)} . \tag{3}$$

Here, we use natural units $\hbar = 2m = 1$. Making $p \to 1$, we recover the Bargmann bound

$$n \leqslant \frac{1}{(2\ell+1)} \int_0^\infty r\,|V_-|\,dr . \tag{4}$$

Varying p in its allowed range, and taking the optimal (lowest) value of the r.h.s. of (1), one usually gets very good agreement (to within 1 %) with the exact results.

On the other hand, it has been shown by Calogero, using Riccati equation [2], that for a spherically symmetric potential which is everywhere attractive, and increasing ($V' \geqslant 0$), one has, for the S-wave ($\ell = 0$), the upper bound

$$n_0 \leqslant \frac{2}{\pi} \int_0^\infty (-V)^{1/2} \, dr \tag{5}$$

and similar bounds for higher waves [3]. Here, the increasing character of the potential is essential, as can be shown by counter-examples. Moreover, for large values of the coupling constant by which we may multiply the potential, it has been shown by several authors [4, 5, 6], that one has, in general, the asymptotic behaviour

$$n_\ell(gV) \underset{g \to +\infty}{\approx} \frac{1}{\pi} g^{1/2} \int_0^\infty |V_-|^{1/2} \, dr \quad . \tag{6}$$

In the present work, we shall fill the gap between (1), which behaves as g^p, and (5) ; that is, we shall derive, for attractive potentials, upper bounds similar to (1), with $1/2 \leqslant p \leqslant 1$. As we shall see, an extra term will appear, the origin of which will be discussed later. As expected, the new bounds become identical to (4) for $p = 1$, and to (5) for $p = \frac{1}{2}$. Obviously, since the Calogero bound is valid under the hypothesis that $V' \geqslant 0$, we must impose some condition on V, besides being attractive everywhere. This extra condition turns out to be

$$\frac{d}{dr}\left[r^{2p-1} (-V)^{p-1}\right] > 0 \quad . \tag{7}$$

To obtain our bounds, we follow the same line of reasoning as for the Calogero bound, which is as follows. Let φ be the reduced radial wave-function for zero energy, and $\ell = 0$ (for simplicity !) :

$$\varphi'' = V(r)\, \varphi(r) \quad , \qquad \varphi(0) = 0 \tag{8}$$

where the potential satisfies the usual Levinson–Jost–Kohn condition

$$\int_0^\infty r \, |V| \, dr < \infty \quad . \tag{9}$$

Writing $(\varphi/\varphi') = r + A$, we get the Riccati equation

$$A' = - V(r + A)^2 \quad , \qquad A(0) = 0 \quad . \tag{10}$$

As is obvious, the function A has poles and zeros which correspond, respectively, to the zeros of φ' and φ, and these are known to be isolated, and interlaced. Denoting the poles by p, and the zeros by z, we have $z_1(=0) < p_1 < z_2 < p_2 < \dots < z_n < p_n < z_{n+1}$, and, eventually, $p_{n+1} = \infty$, which corresponds to $\varphi(\infty) = \alpha(\neq 0)$, and $\varphi'(\infty) = 0$, i.e. an almost bound state at zero energy. In any case, the number of bound states is equal to the number of poles (see figure)

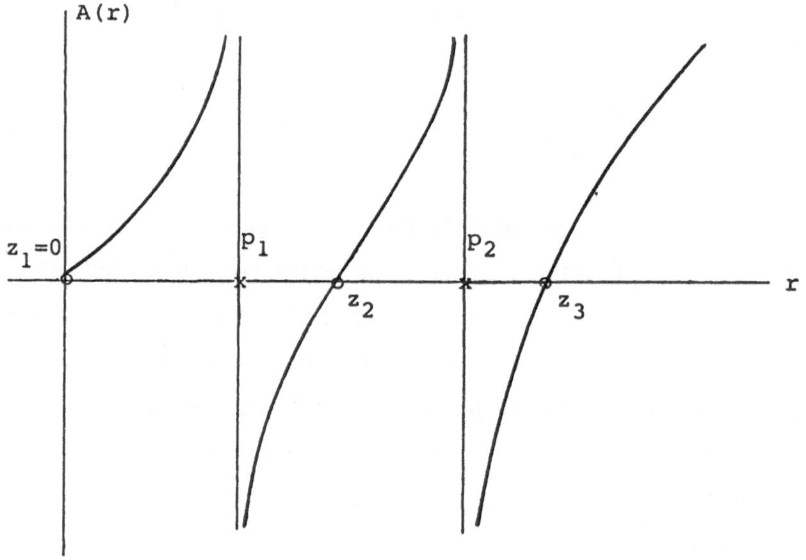

We make now the transformation

$$r + A = r^{2p-1} (-V)^{p-1} \operatorname{tg} \nu \tag{11}$$

with $\frac{1}{2} \leqslant p \leqslant 1$. Whatever p is, the factor in front of tg ν has the dimensionality of r. Using (11) in (10), we get

$$\nu' + \left(\frac{2p-1}{r} + (p-1)\frac{V'}{V} \right) \sin\nu \cos\nu =$$

$$r^{1-2p} (-V)^{1-p} \cos^2\nu + r^{2p-1} (-V)^p \sin^2\nu \quad . \tag{12}$$

The condition to be imposed on V is (7), i.e.

$$\Delta(r) = \frac{2p-1}{r} + (p-1)\frac{V'}{V} > 0 \tag{13}$$

which reduces to $V' \geqslant 0$ for $p = \frac{1}{2}$. If we write now (12) as

$$\nu' + \Delta \sin\nu \cos\nu = F(r)\cos^2\nu + G(r)\sin^2\nu \tag{14}$$

$$F(r) = r^{1-2p}(-V)^{1-p}, \quad G(r) = r^{2p-1}(-V)^p, \tag{15}$$

introduce $1 = 2 - 2p + 2p - 1$ in front of $\cos^2\nu$, and define ω_p by

$$\operatorname{tg}\omega_p = \sup_r \frac{(2p-1)F(r)}{\Delta(r)} \tag{16}$$

we get the inequality

$$\nu' + \frac{\Delta}{\cos\omega_p}\sin(\nu - \omega_p)\cos\nu \leqslant 2(1-p)F(r)\cos^2\nu + G(r)\sin^2\nu \tag{17}$$

which we must integrate now, as has been done for the Calogero bound. Before doing this, we note first that since $\lim r^2V(r) = 0$ as $r \to \infty$, $\operatorname{tg}\omega_p$ is finite, and therefore we can always define

$$0 \leqslant \omega_p < \frac{\pi}{2} . \tag{18}$$

Also, we have $\omega_{1/2} = 0$ and $\omega_1 = \pi/4$, and these correspond to Calogero or Bargmann bounds, respectively, as we shall see below. At any rate, in the ℓ.h.s. of (17), we have $\sin(\nu - \omega_p)\cos\nu \geqslant 0$ if we choose $m\pi + \omega_p \leqslant \nu \leqslant (m + \frac{1}{2})\pi$, $m = 0, 1, \ldots n$. In these ranges of ν, we can therefore neglect this term in (17), and keep the inequality. We write now $\cos^2\nu = \cos^2\nu(\cos^2 2\omega_p + \sin^2 2\omega_p)$ and introduce the greatest of the two functions :

$$H(r) = \operatorname{Sup}[2(1-p)F(r),\ G(r)] . \tag{19}$$

We get then for the above ranges of ν :

$$\nu' \leqslant H(r)(\cos^2 2\omega_p \cos^2\nu + \sin^2\nu) + 2(1-p)F(r)\sin^2 2\omega_p \cos^2\nu . \tag{20}$$

We note now that $\cos^2\nu/(\cos^2 2\omega_p\cos^2\nu + \sin^2\nu)$ is a decreasing function of ν. Therefore, we can replace ν by ω_p in the second term in the r.h.s. of (20). So, finally, we obtain

$$\frac{\nu'}{\sin^2\nu + \cos^2 2\omega_p \cos^2\nu} \leqslant H(r) + \frac{2(1-p)\sin^2 2\omega_p \cos^2\omega_p}{\cos^2\omega_p \cos^2 2\omega_p + \sin^2\omega_p} F(r) . \qquad (21)$$

Integrating (21) in each interval, and summing up, we finally obtain the desired result :

$$n \leqslant C_p \left[\int_0^\infty H(r)\,dr + \frac{2(1-p)\sin^2 2\,\omega_p \cos^2\omega_p}{1 - 4\sin^2\omega_p \cos^4\omega_p} \int_0^\infty F(r)\,dr \right] , \qquad (22)$$

$$C_p = \left[\int_{\omega_p}^{\pi/2} \frac{d\nu}{(\sin^2\nu + \cos^2 2\omega_p \cos^2\nu)} \right]^{-1} \qquad (23)$$

Remarks

1 - It is easily seen that for $p = 1/2$, we get the Calogero bound, and for $p = 1$ the Bargmann bound. Both are obvious from the beginning.

2 - The extra-term in (22), which vanishes at $p = 1/2$ and $p = 1$, has a geometrical origin. Indeed, in the r.h.s. of (20), we are dealing with an ellipse described by the eccentric angle ν, and, in essence, we are using some geometrical properties to go further. For $p = 1/2$, the ellipse becomes a circle, and for $p = 1$, a line, and in both cases, the geometry simplifies.

3 - One can generalize the above results to higher waves, with, of course, more complicated formulae.

4 - Extension of the present method to $p > 1$ may be possible. As is obvious, we get here some infinite terms in the r.h.s. of the inequalities, and the line of reasoning must be modified. All the above points will be dealt with in a forthcoming paper.

In conclusion, it is a great pleasure to dedicate this paper to André Martin on the occasion of the sixtieth anniversary of his birthday. His friendship with one of us (K.C.) from our beginnings in research at the Ecole Normale has been always most valuable and inspiring.

References

[1] Glaser, V., Grosse, H., Martin, A., Thirring, W. : in "Essays in Honor of Valentine Bargmann" /1976/, edited by E.H. Lieb, B. Simon, and A.S. Wightman (Princeton University Press, Princeton).

[2] Calogero, F. /1965/ : Comm. Math. Phys. 1, 80.

[3] Calogero, F. /1967/ : The Variable Phase Approach to Potential Scattering (Academic Press, New York).

[4] Chadan, K. /1968/ : Nuovo Cimento, 58A, 191.

[5] Frank, W. M. /1967/ : J. Math. Phys. 8, 466.

[6] Martin, A. /1972/ : Helv. Phys. Acta 45, 142.

Moment Inequalities and a Martin Conjecture

A.K. Common

Mathematics Institute, University of Kent,
Canterbury CT2 7NF, UK

1 Introduction

It is a great pleasure to have the opportunity to contribute to this volume in honour of André Martin .I first met André twenty years ago when he was working on *rigorous properties* of scattering amplitudes.He encouraged me to publish some extensions I made on his work concerning rigorous properties of $\pi^0\pi^0 \rightarrow \pi^0\pi^0$,s-wave scattering amplitudes /Common,1968/.This initial contact developed into a very happy and fruitful collaboration durng stays of various lengths at C.E.R.N. André's idea of the level of rigour necessary to prove results which are both mathematically and physically interesting is very much in accord with my views.

During the seventies,immediately after the discovery of the charmonium and upsilon systems,André investigated with a number of collaborators properties of quarkonium systems which were not so dependent on a particular model quark-antiquark potential,but more on the potentials general shape.He sent me preprints of his work and again encouraged me to pursue research in this area of *rigorous properties* of potential models.Once again this has proved a very fruitful field of resarch for me.

2 A Conjecture and its Proof

In an early review of the subject/Grosse and Martin,1980/,André made a conjecture concerning the behaviour of energy levels $E(n,\ell)$ as a function of angular momentum ℓ. For ground states $(n = 0)$ it is easy to prove using variational arguments that for any central potential $V(r)$,$E(0,\ell)$ is a concave function of $\ell(\ell+1)$ so that for example:

$$E(0,1) \geq \frac{1}{3}E(0,2) + \frac{2}{3}E(0,0). \qquad (2.1)$$

André wanted to prove a stronger result by restricting the class of potentials and conjectured the following:

Theorem 1

If

$$\frac{d}{dr}[\frac{1}{r}\frac{d}{dr}V(r)] \gtrless 0, \; r > 0, \qquad (2.2)$$

then

$$\frac{\partial^2 E(0,\ell)}{\partial \ell^2} \gtrless 0. \tag{2.3}$$

Note that equality holds in both(2.2)and(2.3) for the harmonic oscillator potential.This conjecture implies that,

$$E(0,1) \gtrless \frac{1}{2}[E(0,2)+E(0,0)] \tag{2.4}$$

instead of the weaker bound (2.1).As for several other conjectures made by André in this field ,supporting arguments were always given e.g. the result given by TH.1. holds at the level of first order perturbation theory.

It was only last year that we were able to give a complete proof of this theorem/ Common and Martin,1988and1989a/.It came from a rather unexpected scource which I will now discuss.Perhaps the most important and beautiful of all the properties proved by André and his co-workers in this field of potential models is given by the following theorem/Baumgartner, Grosse and Martin,1984 and 1985/:

Theorem 2

If

$$\frac{d}{dr}[r^2\frac{dV}{dr}] \gtrless 0,\ r>0, \tag{2.5}$$

then

$$E(n,\ell) \gtrless E(n-1,\ell+1). \tag{2.6}$$

It says how the degeneracy in the coulomb case corresponding to equality in (2.5),(2.6) is split when the potential is changed.

The main ingredient in the proof is the following result:

Lemma

If V(r) satisfies the condition(2.5),then

$$-[\frac{u'_{0,\ell}}{u_{0,\ell}}]' \gtrless \frac{\ell+1}{r^2},\ r>0, \tag{2.7}$$

where $u_{0,\ell}(r)$ is the reduced waved function for the ground state of angular momentum ℓ.

From previous experience,I knew that one could often get interesting results by multiplying an inequality such as (2.7) by a non-negative function f(r) and integrating.For example with $f(r) \equiv u^2_{0,\ell}(r)$,I obtained for potentials satisfying the upper inequality (2.5) /Common,1985/,

$$<T>_{0,\ell} \geq \frac{1}{2}(\ell+1)(2\ell+1)<r^{-2}>_{0,\ell} = \frac{1}{2}(\ell+1)\frac{\partial E(0,\ell)}{\partial \ell}, \tag{2.8}$$

where $<T>_{0,\ell}$ is the average kinetic energy of the ground state of angular momentum ℓ.Similarly by taking

$$f(r) = \int_0^r u^2_{0,\ell}(r')[r'^k <r^n>_{0,\ell} - r'^n <r^k>_{0,\ell}]dr';\ n>k, \tag{2.9}$$

I obtained

$$(2\ell+2+k) < r^{k-1} >_{0,\ell} < r^n >_{0,\ell} \leq (2\ell+2+n) < r^{n-1} >_{0,\ell} < r^k >_{0,\ell} . \quad (2.10)$$

This *reverses* the usual moment inequalities.An improvement on these usual inequalities is obtained when the lower inequality (2.7) is used.

During one of my visits to C.E.R.N. in the spring of 1986 to work with André ,we extended these results to other classes of potentials/Common and Martin ,1987/.For example:

Theorem 3

If

$$\frac{d}{dr}[\frac{1}{r}\frac{dV}{dr}] \gtrless 0, \; r > 0, \quad (2.11)$$

then

$$< T >_{0,\ell} \gtrless \frac{1}{2}(\ell + \frac{3}{2})\partial E(0,\ell)/\partial \ell \quad (2.12)$$

and for $n > k$,

$$(2\ell + 3 + 2n) < r^{2+2k} >_{0,\ell} < r^{2n} >_{0,\ell} \gtrless (2\ell + 3 + 2k) < r^{2k} >_{0,\ell} < r^{2+2n} >_{0,\ell} . \quad (2.13)$$

It is the harmonic oscillator potential which gives equality in (2.12),(2.13). We were able to use (2.12) and similar inequalities to obtain some very nice bounds on grounds state energies in the case of power law potentials as for example:

Theorem 4

For $V(r) = r^\nu, 0 \leq \nu \leq 2$,

$$[(\nu+2)/\nu](\nu/2)^{2/(\nu+2)}[\ell + (\nu+4)/4]^{2\nu/(\nu+2)} \leq E(0,\ell) \leq [(\nu+2)/\nu](\nu/2)^{2/(\nu+2)}[\ell + 3/2]^{2\nu/(\nu+2)}. \quad (2.14)$$

Several months after my return to Canterbury,André wrote to me that he had been able to use these and similar inequalities to prove the conjectured Th.1. for simple power law potentials.There then followed a period of collaborative research at a distance by old fsshioned letter mail,where we were able to extend by stages ,the class of potentials for which the conjecture held.However it was not until the following spring that we discovered that the moment inequalities (2.13) ,which at first did not appear to be particularly useful,held the key to a full proof of Th.1. /Common and Martin,1988/.They were used in a limiting form suggested by André that one gets by taking $n = k + \delta$ and then letting $\delta \to 0_+$:

$$(2\ell + 3 + \beta)^{-1} + \frac{< r^\beta \log r >_{0,\ell}}{< r^\beta >_{0,\ell}} \gtrless \frac{< r^{\beta+2} \log r >_{0,\ell}}{< r^{\beta+2} >_{0,\ell}} \quad (2.15)$$

for any real β for which all moments concerned exist.

Further applications of the above moment inequalities have been the derivation of new constraints on total and kinetic energies of ground states for various classes of potentials such as those defined by (2.2) and (2.5)/Common and Martin,1989a/. Very recently we used generalisations of these inequalities to obtain bounds on the wave function at the origin for these states simply in terms of the corresponding kinetic energies /Common and Martin,1990/.

3 New Moment Inequalities

When we first realised how useful the moment inequalities were,André pointed out that they could be generalised as shown by the following result:

Theorem 5

Let $F(r), G(r)$ be non-negative for $r \geq 0$ and such that $F(r)/G(r)$ is monotonic increasing with r.If

$$\frac{d}{dr}[r^2\frac{dV}{dr}] \gtrless 0,\ r > 0, \tag{3.1}$$

then

$$< F' + [2(\ell+1)/r]F >_{0,\ell} < G >_{0,\ell} \gtrless < G' + [2(\ell+1)/r]G >_{0,\ell} < F >_{0,\ell}, \tag{3.2}$$

where $< F >_{0,\ell} = \int_0^\infty F(r)u_{0,\ell}^2(r)dr$,e.t.c.

Proof

This result is obtained in exactly the same way as the moment inequalities (2.10) except here we take,

$$f(r) \equiv \int_0^r u_{0,\ell}^2(r')[G(r') < F >_{0,\ell} - F(r') < G >_{0,\ell}]dr'. \tag{3.3}$$

As a consequence of this ,we now prove some new inequalities:

Corollary

Define,

$$\begin{aligned} \mu_{k,n} = \pm\Gamma(-k)[-(k+2(\ell+1)) < r^{k-1} >_{0,\ell} < r^n >_{0,\ell} + \\ +(n+2(\ell+1)) < r^k >_{0,\ell} + < r^{n-1} >_{0,\ell}] \end{aligned} \tag{3.4}$$

with $0 > k > -(2\ell+2), n > 0$.

If

$$\frac{d}{dr}[r^2\frac{dV}{dr}] \gtrless 0,\ r > 0, \tag{3.5}$$

then there exists $\phi_n(r) \geq 0; r > 0$, such that

$$\mu_{k,n} = \int_0^\infty t^{-k}\phi_n(t)dt \tag{3.6}$$

Proof

Take $F(r) \equiv r^n, G(r) \equiv e^{-tr}$ in Th.5. whose requirements they satisfy for $t \geq 0$. Then substituting in (3.2),

$$I_n(t) \equiv < te^{-tr} >_{0,\ell} < r^n >_{0,\ell} + n < r^{n-1} >_{0,\ell} < e^{-tr} >_{0,\ell}$$
$$-2(\ell+1)[< e^{-tr}/r >_{0,\ell} < r^n >_{0,\ell} - < e^{-tr} >_{0,\ell} < r^{n-1} >_{0,\ell}]$$
$$\gtrless 0, \; t \geq 0. \tag{3.7}$$

It is straightforward to prove that

$$\mu_{k,n} = \pm \int_0^\infty t^{-(k+1)} I_n(t) dt, \; k > -(2\ell+2), \; n > 0. \tag{3.8}$$

This is done by interchanging the order of integration over t,r and the restrictions on k,n ensure that all integrals concerned exist.The result then is a consequence of using (3.7) in (3.8).

The representation (3.6) ensures that the $\mu_{k,n}$ satisfy the usual moment inequalities/Shohat and Tamarkin,1960/ for fixed n.For example,

$$\mu_{k+\delta,n}\mu_{k-\delta,n} \geq \mu_{k,n}^2 \tag{3.9}$$

for any $\delta > 0$ for which the corresponding moments exist.

From their definition (3.4),it is very interesting to see that the $\mu_{k,n}$ are proportional to the difference between the left and right sides in our original moment inequalities (2.10).Although it would seem that this corollary has no useful application,as in the past we could be very mistaken.

I hope the above discussion has demonstrated the tremendous influence André has had on my research career.May I end by wishing him many more years of fruitful research and hoping that he continue to make conjectures which invariably turn out to be correct.

Acknowledgement

I would like to thank the Theory Division,C.E.R.N. for giving me the opportunity to visit and work with André on many occasions.

References

Baumgartner,B.,H.Grosse,A.Martin/1984/:Phys.Lett.**146B**,363 (Sec.2)

Baumgartner,B.,H.Grosse,A.Martin/1985/:Nucl.Phys.**B254**,528. (Sec.2)

Common,A.K./1968/:Nuovo Cimento **53A**,946. (Sec.1)

Common,A.K./1985/:Journ.Phys.**A18**,2219. (Sec.2)

Common,A.K. and A.Martin/1987/:Journ.Phys.**A20**,4247. (Sec.2)

Common,A.K. and A.Martin/1988/:Euro-Phys.Lett.**4**,1349. (Sec.2)

Common,A.K. and A.Martin/1989a/:Journ.Math.Phys.**30**,601. (Sec.2)

Common,A.K. and A.Martin/1990/:*Preservation of Logarithmic Concavity by the Mellin Transform and Applications to the Schroedinger equation for certain classes of Potentials* : C.E.R.N. preprint in preparation. (Sec.2)

Grosse,H. and A.Martin/1980/:Phys.Rep.**60**,341. (Sec.2)

Shohat,J.A. and J.D.Tamarkin/1960/:The Problem of Moments,(American Math.Soc.,Providence,R.I.). (Sec.3)

On a Class of Spectral Problems Admitting Partial Algebraization

G. Auberson

Laboratoire de Physique Mathématique*, Université Montpellier II, Sciences et Techniques du Languedoc, F-34095 Montpellier Cedex 02, France

Abstract:

Shifman's approach to one-dimensional, spectral problems admitting partial algebraization is studied. The class of Schrödinger operators with a rational potential which are quasi-exactly-solvable by this method is worked out, and is found to reduce to a 4-parameter family of "screened Coulomb potentials".

1. Introduction

Recently, a new class of spectral problems for the Schrödinger operator $H = -\frac{1}{2}\, d^2/dx^2 + V(x)$ has been discovered, which can be *partially solved* by representing H as a quadratic element of the universal enveloping algebra of su(2) /Turbiner, 1988 ab ; Shifman, 1988 a/. Concretely, this means that a *finite* subset of eigenvalues and eigenfunctions (the number of which is related to some integer appearing in H) can be determined *algebraically*, because part of the spectral analysis of H reduces to an eigenvalue problem in the finite dimensional space carrying an irreducible representation of su(2). The class of these so-called "quasi-exactly-solvable" problems has been described exhaustively /Turbiner, 1988 a ; Shifman, 1988 a/. More recently, an extension of the procedure was devised by Shifman, who also produced an example showing that the modified method is effective at least in one case /Shifman, 1988 b/. Still, the power of this method as a device for generating new quasi-exactly-solvable problems remains to be investigated. Such is the purpose of the present paper. Although this modest contribution has no direct bearing on the work of André Martin, it is may be not inappropriate to report on a problem in non relativistic quantum mechanics on the occasion of his 60th birthday. The remarkable advances made in this field by André Martin (from scattering theory to spacing and order of energy levels, not to mention stability of matter) have been a constant source of inspiration for many people.

In Sect. 2 below, the original algebraization procedure for one--dimensional Schrödinger operators is sketched, together with the

---

* Unité de Recherche Associée au CNRS URA 768 (n°040768)

new variant (the reader is referred to previous publications /Turbiner, 1988 a ; Shifman, 1988 ab/ for more detailed accounts). In Sect. 3 I will show how one can try to determine the class C of *all* potentials V(x) for which Shifman's trick applies. This can be done in a systematic way, provided that some statement is true. As I have been unable so far to prove this statement in full generality, it will be left as a conjecture. The construction of C is then straightforward, but quite lenghty. I have only worked out all the *rational* potentials contained in C (Sect. 4). This subclass comes out, rather surprisingly, as a very narrow one, and the same is likely to happen for the whole class C (Sect. 5).

2. Algebraization

One considers the eigenvalue problem

$$H\,\varphi(x) = E\,\varphi(x) \tag{2.1}$$

for the Schrödinger operator

$$H = -\frac{1}{2}\frac{d^2}{dx^2} + V(x) \tag{2.2}$$

on the line $(-\infty, \infty)$ or the half-line $(0,\infty)$.
The preliminary step towards partial algebraization consists in a change of function followed by a change of variable. The change of function is accomplished by a pseudo-gauge transformation

$$\varphi(x) \to \tilde{\varphi}(x) = e^{\alpha(x)}\,\varphi(x) \quad , \tag{2.3}$$

where $\alpha(x)$ is an "arbitrary" real function.
Equations (2.1,2) become :

$$\tilde{H}\,\tilde{\varphi}(x) = E\,\tilde{\varphi}(x) \quad , \tag{2.4}$$

$$\tilde{H} = -\frac{1}{2}\frac{d^2}{dx^2} + A(x)\frac{d}{dx} + V(x) + \frac{1}{2}\,[A'(x) - A^2(x)] \quad , \tag{2.5}$$

where A(x) is the usual "gauge potential"

$$A(x) = \alpha'(x) \quad . \tag{2.6}$$

A change of variable $x \to y(x)$ then transforms the differential operator (2.5) into

$$\tilde{H} = P_4(y)\frac{d^2}{dy^2} + P_3(y)\frac{d}{dy} + P_2(y) \quad , \tag{2.7}$$

where :

$$\begin{cases} P_4(y) = -\dfrac{1}{2}\left(\dfrac{dy}{dx}\right)^2 \quad , \\ P_3(y) = A(x)\,\dfrac{dy}{dx} - \dfrac{1}{2}\,\dfrac{d^2y}{dx^2} \quad , \\ P_2(y) = V(x) + \dfrac{1}{2}\,[A'(x) - A^2(x)] \quad . \end{cases} \tag{2.8}$$

So far, the function $y(x)$ is arbitrary, but the notation anticipates the fact that the P_n's are forced to be polynomials of degree n in y. This is because one demands that $\tilde{H}$ be expressible in terms of su(2) elements. To this end, one can use any $(2j+1)$-dimensional, irreducible representation of the su(2) algebra, with generators in differential form /Vilenkin, 1968/ :

$$\begin{cases} J_+ = 2jy - y^2\,\dfrac{d}{dy} \quad , \\ J_0 = -\,j + y\,\dfrac{d}{dy} \quad , \\ J_- = \dfrac{d}{dy} \quad . \end{cases} \tag{2.9}$$

These generators obey the su(2) commutation rules :

$$[J_0, J_+] = J_+ \quad , \quad [J_0, J_-] = -\,J_- \quad , \quad [J_+, J_-] = 2J_0 \quad , \tag{2.10}$$

as well as :

$$J_+J_- = -\,J_0^2 + J_0 + j(j+1) \quad . \tag{2.11}$$

The linear space carrying this representation is the space of polynomials of degree $2j$ in y.

Clearly, if $\tilde{H}$ can be written as a (quadratic) expression h in J_+, J_0 and J_-, a set of $(2j+1)$ eigenvalues E_n and eigenfunctions $\phi_n(y)$ (= polynomials of degree $2j$) will be obtained algebraically for $\tilde{H} = h$:

$$h\,\phi_n(y) = E_n\,\phi_n(y) \qquad (n = 0, 1, \ldots, 2j) \quad , \tag{2.12}$$

and then, by reversing the previous steps, for H. The most general expression of h is

$$h = C_{++}J_+^2 + C_{+0}J_+J_0 + C_{00}J_0^2 + C_{0-}J_0J_- + C_{--}J_-^2 + C_+J_+ + C_0J_0 + C_-J_- + b \tag{2.13}$$

where $C_{\alpha\beta}$, C_α, b are arbitrary real coefficients (Equations (2.10,11) have been used to reduce the number of terms in (2.13)). The identification

$$\tilde{H} = h \tag{2.14}$$

then leads, via (2.9), to the following expressions of the polynomials P_n :

$$\begin{cases} P_4(y) = C_{++}y^4 - C_{+0}y^3 + C_{00}y^2 + C_{0-}y + C_{--} \quad , \\ P_3(y) = -2(2j-1)C_{++}y^3+[(3j-1)C_{+0}-C_+]y^2+[-(2j-1)C_{00}+C_0]y-jC_{0-}+C_- , \\ P_2(y) = 2j(2j-1)\ C_{++}y^2 + 2j\ (-jC_{+0} + C_+)y + j^2C_{00} - j\ C_0 + b\ . \end{cases} \tag{2.15}$$

Given h (i.e. the set of coefficients $C_{\alpha\beta}, C_\alpha$ and b), H is now reconstructed by solving the equations (2.8) for V(x) :

$$\begin{cases} x = \int^y dz \dfrac{1}{\sqrt{-2\ P_4(z)}} \quad \text{(giving y as a function of x)} \quad , \\ A(x) = \dfrac{P_3(y) - \frac{1}{2} P'_4(y)}{\sqrt{-2\ P_4(y)}} \quad , \\ V(x) = P_2(y) + \dfrac{1}{2}[A^2(x) - A'(x)] \quad . \end{cases} \tag{2.16}$$

Equations (2.15,16) define the general class of potentials V(x) (depending on the half-integer j) for which the spectral problem (2.1) is quasi-exactly-solvable. That is, part of the point spectrum of H is given by

$$H\ \varphi_n(x) = E_n\ \varphi_n(x) \quad \text{with} \quad \varphi_n(x) = e^{-\alpha(x)}\ \phi_n(y(x)) \quad . \tag{2.17}$$

Or course, one has to retain only those solutions of (2.12) for which the corresponding eigenfunctions $\varphi_n(x)$ are normalizable and belong to the (properly defined) domain of H.

The full catalogue of such su(2)-based potentials can be found in the literature (e.g. they are listed in /Turbiner, 1988 a/ under the name of "first kind" potentials). I will here simply conform to the common practice of quoting the simplest example :

$$V(x) = -\frac{1}{2}(8j+3)x^2 + \frac{1}{2}x^6 \quad \left(x \in (-\infty,\infty);\ j = 0, \frac{1}{2}, 1, \frac{3}{2}, \ldots\right) \ . \tag{2.18}$$

Let me now describe Shifman's variant (which I have slightly generalized here[1]).

---

1) In /Shifman, 1988 b/, the functions G and f introduced below are identified.

One first introduces a real parameter "a" and represents the "gauge potential" $A(x)$ as

$$A(x\ ;\ a) = a + G(x) \quad , \tag{2.19}$$

with a function G *not depending on* a. The gauge transformed Hamiltonian (2.5) becomes

$$\tilde{H} = -\frac{1}{2}\frac{d^2}{dx^2} + [a + G(x)]\frac{d}{dx} + \mathcal{V}(x) - a\,G(x) - \frac{1}{2}a^2 \quad , \tag{2.20}$$

where

$$\mathcal{V}(x) = V(x) + \frac{1}{2}[G'(x) - G^2(x)] \quad . \tag{2.21}$$

One next introduces an "arbitrary" real function $f(x)$, *not depending on* a, and imposes the identity

$$\frac{1}{f(x)}\left(\tilde{H} + \frac{1}{2}a^2\right) = h \tag{2.22}$$

generalizing (2.14). Then, a simple calculation shows that the equations (2.16) are replaced by :

$$\begin{cases} \text{a)} \quad -\dfrac{1}{2f(x)}\left(\dfrac{dy}{dx}\right)^2 = P_4(y) \quad , \\ \text{b)} \quad \left[\dfrac{a + G(x)}{f(x)} - \dfrac{f'(x)}{4\,f^2(x)}\right]\left(\dfrac{dy}{dx}\right) = P_3(y) - \dfrac{1}{2}P'_4(y) \quad , \\ \text{c)} \quad \dfrac{1}{f(x)}[\mathcal{V}(x) - a\,G(x)] = P_2(y) \quad , \end{cases} \tag{2.23}$$

where the polynomials P_2, P_3, P_4 are still defined by (2.15).

The method then proceeds as follows. Given a function $f(x)$ with suitable "boundary" behaviour (at $x \to \infty$ and $x \to -\infty$ or 0), one can solve the differential equation (2.23a) for y, which produces a function $y(x\ ;\ a)$. Equations (2.23 b,c) then successively fix $G(x)$ and $\mathcal{V}(x)$, hence $V(x)$. It is important to notice however that the coefficients $C_{\alpha\beta}$, C_α, b (some of which necessarily depend on a), as well as the function $f(x)$, must be so chosen as to make the solutions $G(x)$ and $\mathcal{V}(x)$ not depending on a. These constraints, to be discussed in Sect. 3, severely limit the flexibility of the procedure.

Suppose now that, given $j = 0, \frac{1}{2}, 1, \ldots,$ the finite-dimensional eigenvalue problem

$$h\,\phi_n(y\ ;\ a) = \epsilon_n(a)\,\phi_n(y\ ;\ a) \qquad (n = 0,1,\ldots,2j) \tag{2.24}$$

has been solved (algebraically). This means that

$$\left[\tilde{H} - \epsilon_n(a)\,f(x)\right]\phi_n(y\ ;\ a) = -\frac{a^2}{2}\,\phi_n(y\ ;\ a) \quad , \tag{2.25}$$

or, coming back to the operator H, that

$$\left[-\frac{1}{2}\frac{d^2}{dx^2} + V(x) - \epsilon_n(a)\ f(x)\right] \varphi_n(x\ ;\ a) = -\frac{a^2}{2}\varphi_n(x\ ;\ a) \quad (2.26)$$

with

$$\varphi_n(x\ ;\ a) = \exp\left[-\ ax - \int^x dz\ G(z)\right]\ \phi_n(y(x\ ;\ a);\ a) \quad . \quad (2.27)$$

Setting

$$\epsilon_n(a) = c = \text{given, n } \textit{independent} \text{ number} \quad , \quad (2.28)$$

one obtains by inversion a collection of values a_n for "a" which partially solve the problem (2.1) for the potential

$$\bar{V}(x) = V(x) - cf(x) = \mathcal{V}(x) + \frac{1}{2}[G^2(x) - G'(x)] - cf(x) \quad . \quad (2.29)$$

A set of eigenvalues and eigenfunctions is produced :

$$E_n = -\frac{1}{2}a_n^2 \quad , \quad \varphi_n(x) = \varphi_n(x\ ;\ a_n) \quad . \quad (2.30)$$

It remains to determine the class of potentials $\bar{V}(x)$ which can be reached by this procedure. At first sight, such a task appears as quite complicated. The argument of the next section will make it feasible.

3. A conjecture

Clearly, the formalism of Sect. 2 is invariant with respect to, both, affine (a-independent) transformations of the variable x and affine (a-dependent) transformations of the variable y. This freedom leads to noticeable simplifications in (2.23). In fact, I claim that it allows us to choose the variable y as independent of the parameter "a". To see how this comes about, consider a particularly simple case where some of the coefficients $C_{\alpha\beta}$, C_α identically vanish (remember that these are in general functions of a) :

$$C_{++} = C_{+0} = C_{00} \equiv 0 \quad , \quad C_{0-} \not\equiv 0 \quad ; \quad C_+ \equiv 0 \quad . \quad (3.1)$$

Equation (2.33a) then reads

$$-\frac{1}{2\ f(x)}\left(\frac{dy}{dx}\right)^2 = C_{0-}(a)y + C_{--}(a) \quad . \quad (3.2)$$

By a suitable affine transformation of y, one can always set :

$$C_{0-}(a) \equiv 1 \quad , \quad C_{--}(a) \equiv 0 \quad , \quad (3.3)$$

in which case the general solution of (3.2) is

$$y(x\ ;\ a) = [F(x) + D(a)]^2 \tag{3.4}$$

with

$$f(x) = -\,2\,F'(x)^2\ . \tag{3.5}$$

Next, (2.23c) becomes

$$\frac{\mathcal{V}(x)}{f(x)} - a\,\frac{G(x)}{f(x)} = -\,j\,C_0(a) + b(a)\ , \tag{3.6}$$

which implies the existence of two constants μ and ν such that :

$$\mathcal{V}(x) = \mu\, f(x)\ , \tag{3.7}$$

$$G(x) = \nu\, f(x)\ , \tag{3.8}$$

$$b(a) = \mu - \nu a + j\,C_0(a)\ . \tag{3.9}$$

Finally, the insertion of (3.4,5,8) into (2.23b) gives

$$\left[2\nu\,F'(x) + \frac{F''(x)}{2F'(x)^2} - \frac{a}{F'(x)}\right][F(x) + D(a)] =$$

$$C_0(a)\,[F(x) + D(a)] + C_-(a) - \left(j + \frac{1}{2}\right)\ . \tag{3.10}$$

From this identity in x and a, one readily infers that $C_0(a)$ and $C_-(a)$ are affine functions and that D(a) is a constant. This means that, with the choice (3.3), y(x ; a) cannot depend on a.
Analogous (but longer !) calculations carried out in several cases lead to the same conclusion, and give strong indication that :

$$\left\{\begin{array}{l}\text{Any function } y(x\ ;\ a) \text{ compatible with the equations (2.33)}\\ \text{is equivalent (modulo some a-dependant affine transformation)}\\ \text{to a function } y(x) \text{ no longer depending on a.}\end{array}\right. \tag{3.11}$$

I have not been able so far to complete the proof of this statement (some cases involve heavy use of elliptic functions), which I leave as a *conjecture*.
As a consequence of (3.11), the structure of the equations (2.23) considerably simplifies. First of all, because of (2.23a), *the coefficients* $C_{\alpha\beta}$ *can be chosen as independent of* a.
Next, (2.23 b,c) assume the form :

$$\left\{\begin{array}{l}F_1(x) + a\,F_2(x) = -\,C_+(a)\,y^2(x) + C_0(a)\,y(x) + C_-(a)\ ,\\ F_3(x) + a\,F_4(x) = 2j\,C_+(a)\,y(x) - j\,C_0(a) + b(a)\ ,\end{array}\right. \tag{3.12}$$

implying that the remaining coefficients are necessarily affine functions of a :

$$C_{\pm}(a) = B_{\pm} + A_{\pm}a \quad , \quad C_0(a) = B_0 + A_0 a \quad , \quad b(a) = b_1 + b_2 a \quad . \tag{3.13}$$

The a-dependence is then explicitly removed from the equations (2.23) which split up into :

$$-\frac{1}{2\ f(x)}\left(\frac{dy}{dx}\right)^2 = P_4(y) \quad , \tag{3.14}$$

$$\frac{1}{f(x)}\left(\frac{dy}{dx}\right) = Q_2(y) \quad , \tag{3.15}$$

$$\left[g(x) - \frac{f'(x)}{4\ f^2(x)}\right]\left(\frac{dy}{dx}\right) = Q_3(y) \quad , \tag{3.16}$$

$$g(x) = Q_1(y) \quad , \tag{3.17}$$

$$v(x) = 2j(2j-1)\ C_{++}y^2 + 2j(B_+ - j\ C_{+0})y + j^2\ C_{00} - j\ B_0 + b_1 \quad , \tag{3.18}$$

where :

$$g(x) = G(x)/f(x) \quad , \quad v(x) = \mathcal{V}(x)/f(x) \tag{3.19}$$

and :

$$\begin{cases} Q_2(y) = -\ A_+y^2 + A_0 y + A_- \quad , \\ Q_3(y) = -4jC_{++}y^3 + \left[\left(3j + \frac{1}{2}\right)C_{+0} - B_+\right]y^2 + (B_0 - 2jC_{00})y - \left(j + \frac{1}{2}\right)C_{0-} + B_- \quad , \\ Q_1(y) = -\ 2j\ A_+y + j\ A_0 - b_2 \quad . \end{cases} \tag{3.20}$$

The elimination of dy/dx, f and g between (3.14-17) generates the *consistency condition*

$$P_4(2Q_1 + Q'_2) = Q_2\left(\frac{1}{2}P'_4 - Q_3\right) \quad . \tag{3.21}$$

Once this condition is satisfied, the equations (3.14-16) can be replaced by :

$$\frac{dy}{dx} = -\ 2\ \frac{P_4(y)}{Q_2(y)} \quad , \tag{3.22}$$

$$f(x) = -\ 2\ \frac{P_4(y)}{Q_2(y)^2} \quad . \tag{3.23}$$

At this point, one has a well-defined procedure to construct the potential from the set of constants $C_{\alpha\beta}$, B_α, A_α, b_1, and b_2 : the functions $y(x)$, $f(x)$, $g(x)$ and $v(x)$ are successively determined by (3.22,23,17,18) ; hence $\bar{V}(x)$ from (3.19) and (2.29). One observes however that the freedom which the method seemed to offer in the beginning was somewhat illusory, since i) the constraint (3.21) put strong restrictions on the choice of $C_{\alpha\beta}$, B_α, A_α, and b_2, ii) the function f is completely fixed by these constants, through (3.23).

4. Solving for rational potentials

From now on, I shall concentrate on the class of rational potentials $\bar{V}(x)$. This will be enough to get some insight into the limitations of the method. According to (3.17-23), the functions f, G, $\mathcal{V}$ and $dG/dx = - 2\, P_4/Q_2 \cdot dG/dy$ can be viewed as rational functions of y. This is also true for the potential $\bar{V}$ given by (2.29). Therefore, requiring that $\bar{V}$ be a rational function of x implies that y(x) and its inverse x(y) are algebraic functions. Then, from (3.22), i.e.

$$x = - \frac{1}{2} \int^y dz \, \frac{Q_2(z)}{P_4(z)} \quad , \tag{4.1}$$

one derives a further restriction :

"The residue of Q_2/P_4 at each pole must vanish" . (4.2)

The task of uncovering the class of all rational potentials which are quasi-exactly-solvable by the present method is now a simple matter of computation. However, there are many cases to examine individually, corresponding to various possibilities of vanishing of certain constants $C_{\alpha\beta}$, B_α, A_α. As an illustration, let me just single out one short example :

$$C_{++} = 0 \quad , \quad C_{+0} \neq 0 \quad . \tag{4.3}$$

By a rescaling of y, one is allowed to set $C_{+0} = 1$. Because of (4.2), at least two zeros of the 3[d] degree polynomial P_4 must coincide. This double zero may be fixed at the origin by a translation of y, so that

$$P_4(y) = - y^2 (y - y_0) \qquad (\text{i.e. } C_{00} = y_0 \ , \ C_{0-} = C_{--} = 0) \quad . \tag{4.4}$$

The condition (4.2) further requires that :

$$Q_2(y_0) = 0 \ , \ Q_2'(0) y_0 + Q_2(0) = 0 \quad (\text{i.e. } A_+ = 0 \ , \ A_- = - A_0 y_0), \tag{4.5}$$

and the concistency condition (3.21) that :

$$\begin{cases} B_0 = (2j+1)y_0 \quad , \quad B_- = 0 \quad , \\ b_2 = \frac{1}{2}\,[B_+ - (j+1)]A_0 \quad . \end{cases} \tag{4.6}$$

Equations (3.22,23,17,18) then give :

$$y = -\frac{A_0}{2x} \quad , \quad f(x) = \frac{-1}{x(A_0 + 2y_0 x)} \quad ,$$

$$g(x) = \frac{1}{2}\,(3j+1 - B_+)A_0 \quad , \quad v(x) = \frac{j(j-B_+)}{x} + b_1 \quad . \tag{4.7}$$

Finally, (3.19) and (2.29) yield :

$$\bar{V}(x) = \frac{(B_+ + j)^2 - 1}{8x^2} + \frac{[B_+ - (3j+1)][B_+ - 3(j+1)]\; y_0^2}{2(A_0 + 2y_0 x)^2}$$

$$+ \frac{(j+1)\;[2\,B_+ - (3j+1)]y_0 - B_+^2 y_0 + 2(c-b_1)}{2x\;(A_0 + 2y_0 x)} \quad . \tag{4.8}$$

Such a potential is non singular, provided that it is restricted to the half-line $x > 0$ and that $A_0 y_0 \geqslant 0$. Then, it may be viewed as an effective, 3-dimensional, central potential in the ℓ-partial wave. The first term is identified with the centrifugal term if one sets $B_+ = -j - (2\ell+1)$. Reminding that $\mathscr{P}_n(y\;;\;a)$ is a polynomial of degree $2j$ in $y = -A_0/2x$ and using (2.27), one can check indeed that the eigenfunctions $\varphi_n(x\;;\;a)$ are normalizable on $(0,\infty)$ for $a > 0$ and exhibit a regular ℓ-wave behaviour at the origin :
$\varphi_n(x\;;\;a) \underset{x\to 0}{\simeq} \text{const.}\; x^{\ell+1}$.

Performing similar calculations in all cases, collecting the results, and ignoring "uninteresting" potentials[2)], one ends up with only *one* family of rational potentials, namely :

$$\bar{V}(x) = \frac{j(j+1) + c}{2(x^2+\sigma)} + \frac{\left[\frac{1}{8}\,B_-^2 - 2j(j+1)\sigma\right] - \left(j + \frac{1}{2}\right)B_- x}{(x^2+\sigma)^2} \quad , \tag{4.9}$$

corresponding to the following assignment of the coefficients :

---

2) By "uninteresting" potentials are meant those for which the eigenfunctions $\varphi_n(x\;;\;a)$ are not normalizable (on the line or the half-line) as well as those which are trivial or exactly solvable, like the Coulomb plus centrifugal potential.

$$\begin{cases} C_{++} = C_{+0} = C_{0-} = 0 \ , \quad C_{00} = 1 \ , \quad C_{--} = \sigma \ , \\ C_+ = 2a \ , \quad C_0 = -(2j+1) \ , \quad C_- = B_- - 2\sigma a \ , \\ b = - B_- a \ , \end{cases} \tag{4.10}$$

and the following expressions of the relevant functions :

$$y(x) = x \ , \ f(x) = \frac{-1}{2(x^2+\sigma)}, \ g(x) = - 4jx + B_- , \ v(x) = j(3j+1). \tag{4.11}$$

For $\sigma > 0$, the potentials (4.9) are not quasi-exactly-solvable on the line $(-\infty, \infty)$ because the eigenfunctions

$$\varphi_n(x \ ; \ a) = (x^2+\sigma)^{-j} \exp\left[-ax - \frac{B_-}{2\sqrt{\sigma}} \operatorname{ctg}^{-1} \frac{x}{\sqrt{\sigma}}\right] \phi_n(x \ ; \ a) \tag{4.12}$$

are not normalizable.

For $\sigma = 0$, one recovers exactly the central potential of /Shifman, 1988 b/ (Equation (2) of this reference).

For $\sigma \equiv - r_0^2/4 < 0$, one obtains a new family of quasi-exactly-solvable, central potentials, which is conveniently described by setting

$$x = r + \frac{r_0}{2} \tag{4.13}$$

and :

$$\begin{cases} B_- = 2(j+\ell+1) \ r_0 \ , \\ c = 2(j+\ell+1)^2 + j(j+1) + 2\alpha r_0 \ . \end{cases} \tag{4.14}$$

The expression (4.9) then becomes

$$\bar{V}(r) = \frac{\ell(\ell+1)}{2r^2} + \alpha\left(\frac{1}{r} - \frac{1}{r+r_0}\right) + \frac{(2j+\ell+1)\ (2j+\ell+2)}{2(r+r_0)^2} \ , \tag{4.15}$$

with eigenfunctions

$$\varphi_n(r \ ; \ a) = \frac{r^{\ell+1}}{(r+r_0)^{2j+\ell+1}} e^{-ar} \phi_n\left(r + \frac{r_0}{2} \ ; \ a\right) \ . \tag{4.16}$$

It is readily checked that the potential previously derived, i.e. (4.8), can also be put in the form (4.15). This is a particular instance of a general fact, namely that different choices of the set of coefficients $C_{\alpha\beta}$, C_α and b may deliver the same potential.

5. Conclusion

To summarize, I have found that (if the conjecture (3.11) is true !) the method of /Shifman, 1988 b/ for generating quasi-exactly-

-solvable spectral problems produces, in the class of rational potentials, the only new result (4.15) (besides the example already given in that reference). This is a family of "screened Coulomb potentials" depending, for each $j = 0, \frac{1}{2}, 1, \ldots,$ on the 3 free parameters r_0, α and ℓ.

As for the remaining class of non rational potentials, the restrictions imposed by (3.23) and the consistency condition (3.21) still apply, and are likely to limit again considerably its extent. It is somewhat surprising that the clever Shifman's trick reveals no more rewarding. May be this is just an indication that a further extension of the original method remains to be discovered.

References

Shifman, M.A. /1988 a/ : "New findings in quantum mechanics", XXVIII Cracow School of Theor. Phys., Zakopane.

Shifman, M.A. /1988 b/ : Bern University preprint BUTP-88/32.

Turbiner, A.V. /1988 a/ : Commun. Math. Phys. 118, 467.

Turbiner, A.V., Ushveridze, A.G. /1988 b/ : Phys. Lett. 162A, 181.

Vilenkin, N.Ys. /1968/ : *Special functions and the theory of group representations* (Transl. of Math. Monographs, Vol. 22), Amer. Math. Soc., Providence.

High Energy Theorems

S.M. Roy and V. Singh

Tata Institute of Fundamental Research, Homi Bhabha Road,
Bombay 400005, India

ABSTRACT

We review analyticity properties, bounds and high energy theorems for strong interaction scattering amplitudes following from the postulates of axiomatic field theory.

1. Introduction. The bound $\sigma_{tot} <$ const $[\ln(s/s_0)]^2$ where s is the square of the c.m. energy, and s_0 a constant, was proved by Froissart /1961/ under the assumption of Mandelstam representation with a finite number of subtractions. This assumption is no longer considered reasonable due to the possibility of infinitely rising Regge trajectories ($\alpha(t) \sim$ const $\times\, t, t =$ c.m. momentum transfer squared) suggested by 'duality'. A more general framework is needed. André Martin /1963a, 1966a/ launched the field of analyticity properties, bounds and high energy theorems for strong interaction scattering amplitudes in axiomatic field theory by his celebrated proof of the Froissart bound (now called Froissart-Martin bound) in this framework. Among his other monumental contributions to this field are:

(i) extension of the axiomatic analyticity domain in t of the scattering amplitude and its absorptive part, the Lehmann-Martin ellipses /Lehmann, 1958, 1959; Martin, 1966a/;

(ii) proof of fixed-t dispersion relations in a domain including positive unphysical t /Martin, 1966a, 1966b; Jin and Martin, 1964/;

(ii) a veritable 'bootstrap' calculation deriving rigorous absolute bounds on the pion-pion amplitudes in terms of pion-mass alone /Martin, 1965a; Lukaszuk and Martin 1967/, (units $m\pi = 1$), $|F(2,2)| < 37, -100 < F(4/3,4/3) < 16$;

(*iv*) a theorem that within the diffraction peak, the ratio of elastic differential cross sections for the processes $AB \to AB$ and $A\bar{B} \to A\bar{B}$ approaches unity for $s \to \infty$ /Cornille and Martin 1972a,b,c; 1974/.

We briefly review the principal results. Earlier reviews may be consulted for further details: /Martin 1969, 1970; Eden 1971; Singh 1971; Roy 1972; Fisher 1981; Valin 1989/. Of course, some new results not included in the earlier reviews will be reported here.

In the simplest case of spinless scattering $AB \to CD$ we denote the scattering amplitude by $F(s, \cos\theta)$, with the partial wave expansion

$$F(s,\cos\theta) = \frac{\sqrt{s}}{2k}\sum_{\ell=0}^{\infty}(2\ell+1)a_\ell(s)P_\ell(\cos\theta), \tag{1}$$

and with normalization specified by

$$\frac{d\sigma}{d\Omega} = |2F(s,\cos\theta)/\sqrt{s}|^2. \tag{2}$$

Here k denotes the initial c.m. momentum and θ the c.m. scattering angle for elastic scattering

$$t = -2k^2(1-\cos\theta). \tag{3}$$

2. <u>Unitarity and Positivity Properties</u>. The interplay between unitarity and analyticity became in the hands of Martin a most powerful tool. Unitarity of the S-matrix implies for an elastic amplitude F

$$\begin{aligned}&\int h^*(\vec{n}_1)ImF(s,\vec{n}_1\cdot\vec{n}_2)h(\vec{n}_2)[d\vec{n}_1/(4\pi)][d\vec{n}_2/(4\pi)]\\ &\geq (2k/\sqrt{s})\int h^*(\vec{n}_1)F^*(s,\vec{n}_1\cdot\vec{n}_3)F(s,\vec{n}_3\cdot\vec{n}_2)h(\vec{n}_2)\prod_{i=1}^{3}[d\vec{n}_i/(4\pi)],\end{aligned} \tag{4}$$

where $h(\vec{n})$ is a square-integrable function of the unit vector $\vec{n}$; or, in terms of partial waves,

$$1 \geq Im a_\ell(s) \geq |a_\ell(s)|^2, \tag{5}$$

which implies the positivity of $Im a_\ell(s)$. Using the relation /see e.g. Martin 1969, p.118/

$$P_\ell(\cos\theta) = \sum_{k=0}^{\ell} g_k g_{\ell-k}\cos(\ell-2k)\theta, \tag{6}$$

where

$$g_k = 2^{-2k}\binom{2k}{k}, \tag{7}$$

we see that

$$ImF(s,\cos\theta) = \sum_{n=0}^{\infty} c_n(s)\cos(n\theta);\ \ c_n(s) \geq 0. \tag{8}$$

A similar property for the derivatives w.r.t. $\cos\theta$ follows from

$$\frac{d}{d\cos\theta}\cos n\theta = n\frac{\sin n\theta}{\sin\theta} = n\left[2\cos(n-1)\theta + \frac{\sin(n-2)\theta}{\sin\theta}\right], \tag{9}$$

which yields

$$\frac{d^n}{d(\cos\theta)^n} ImF(s,\cos\theta) = \sum_{m=0}^{\infty} d_m(s)\cos m\theta;\ \ d_m(s) \geq 0. \tag{10}$$

A consequence of this positivity property exploited widely by Martin is that, for all $n \geq 0$, and $-1 \leq \cos\theta \leq 1$,

$$\left|\frac{d^n}{d(\cos\theta)^n} ImF(s,\cos\theta)\right| \leq \left[\frac{d^n}{d(\cos\theta)^n} ImF(s,\cos\theta)\right]_{\cos\theta=1}. \tag{11}$$

Particles with spin. Cornille and Martin /1976/ showed that Eq. (8) also holds with $ImF(s,\cos\theta)$ replaced by $(d\sigma_A/d\Omega)_{\text{unpolarised,el.}}$ which denotes the contribution of the absorptive part to the unpolarized elastic differential cross section. Mahoux /1976/ proved the more powerful result

$$(d\sigma_A/d\Omega)_{\text{unpolarised,el.}} = \sum_{\ell=0}^{\infty} \sigma_\ell(s) P_\ell(\cos\theta);\ \ \sigma_\ell(s) \geq 0. \tag{12}$$

Inelastic Processes. For $a+b \to a'+b'$ with a, b, a', b' specifying particle types as well as helicities, it has been proved that /Roy 1977/ the absorptive cross section has the representation

$$k_a^2\ (d\sigma_A/d\Omega)_{\vec{n}'a'b';\vec{n}ab} = \sum_{\ell=0}^{\infty} P_\ell(\vec{n}\cdot\vec{n}')[\sigma_\ell(s)]_{a'b';ab} \tag{13}$$

where $\sigma_\ell(s)$ is a matrix with non-negative eigenvalues. Here $\vec{n}, \vec{n}'$ denote the momentum directions of a, a' in the c.m. system, and k_a denotes the magnitude of c.m. momentum of particle a.

3. Analyticity Properties and Dispersion Relations.

3.1 Fixed t dispersion relations, $-t_1 < t \leq 0$.

Dispersion relations for $t = 0$ /Symanzik, 1957/ and for fixed $t < 0$ /Bremermann, Oehme, Taylor, 1958; Bogoliubov, Medvedev and Polivanov, 1958/ have been proved starting from LSZ local field theory for a number of processes $AB \to AB$ such as $\pi N \to \pi N$, $\pi\pi \to \pi\pi$, $\pi\pi \to K\bar{K}$, $\pi K \to \pi K$, $KK \to KK$, $\pi\Lambda \to \pi\Lambda$, $\pi\Sigma \to \pi\Sigma$. The scattering amplitude for $-t_1 \leq t \leq 0$ is the boundary value of a real analytic function of s,

$$F_{AB\to AB}(s,t) = \lim_{\epsilon\to 0+} F(s+i\epsilon, t); \; s \geq (m_A+m_B)^2, \tag{14}$$

where $F(s,t) = F^*(s^*,t)$ is holomorphic in the s-plane except for stable particle poles and cuts along real s for $s \geq s_{\text{threshold}}$, and $s \leq 2m_A^2 + 2m_B^2 - t - u_{\text{threshold}}$. Here $u \equiv 2m_A^2 + 2m_B^2 - t - s$. On the left-hand cut,

$$F_{AB\to AB}(u,t) = \lim_{\epsilon\to 0+} F(s-i\epsilon, t), \; u \geq (m_A+m_B)^2. \tag{15}$$

In addition, for $-t_1 \leq t \leq 0$, $F(s,t)$ is polynomially bounded in LSZ formalism, in Wightmann formalism /Hepp 1964/, in Jaffe's theory, as well as in Araki's theory of local observables /Epstein, Glaser and Martin 1969/. Hence we have the dispersion relation

$$\begin{aligned} F(s,t) = \sum_{n=0}^{N-1} c_n(t)s^n &+ \frac{s^N}{\pi}\int_{s_{\text{thr.}}}^{\infty} \frac{A_s(s',t)}{s'^N(s'-s)}ds' \\ &+ \frac{u^N}{\pi}\int_{u_{\text{thr.}}}^{\infty} \frac{A_u(u',t)}{u'^N(u'-u)}du'. \end{aligned} \tag{16}$$

Here $A_s(s,t)$ and $A_u(u,t)$ are physical absorptive parts in the s- and u-channels respectively if the corresponding $\cos\theta$ is physical; they have to be defined by analytic continuation of the partial wave expansion if $\cos\theta < -1$. E.g. for $t < 0$ and $k^2 \to 0$, $\cos\theta \to -\infty$. For analytic continuation in $\cos\theta$ we make crucial use of the "Lehmann ellipses" /Lehmann 1958, 1959/. Lehmann proved that (i) $F(s, \cos\theta)$ is analytic inside an ellipse in the complex $\cos\theta$-plane with foci ± 1 and semi major axis

$$\cos\theta_0(s) = \left[1 + \frac{(M_A'^2 - M_A^2)(M_B'^2 - M_B^2)}{k^2(s-(M_A' - M_B')^2)}\right]^{1/2}, \tag{17}$$

where M_A' is the mass of the lowest state A' such that $< A'|j_A(0)|0 > \neq 0$, and M_B' is similarly defined (for j_π, $M_A' = 3m_\pi$); (ii) $A_s(s, \cos\theta)$ is analytic inside the large Lehmann-ellipse with foci ± 1 and semimajor axis $2\cos^2\theta_0(s) - 1$.

The large Lehmann-ellipse allows continuation of the partial wave expansion for $A_s(s, \cos\theta)$ to values of $\cos\theta \sim -\text{const}/k^2$ for $k^2 \to 0$, and hence defines $A_s(s,t)$ in fixed negative t dispersion relations. Thus, fixed-t dispersion relations are proved in $-t_1 \leq t \leq 0$ with $t_1 = 28m_\pi^2$ for pion-pion scattering and $t_1 = 12.4m_\pi^2$ for pion-nucleon scattering.

3.2 Dispersion relations for fixed t in complex domain including $|t| < R$.

In addition to the above dispersion relations for $-t_1 \leq t \leq 0$ valid for some processes, Bros, Epstein and Glaser /1964/ proved that for any reaction $A + B \to C + D$ where A, B, C, D are stable particles, any point $s_1, \cos\theta_1$ in the physical region has an analyticity neighbourhood $|s - s_1| < n(s_1, \cos\theta_1)$, $|\cos\theta - \cos\theta_1| < n(s_1, \cos\theta_1)$ where the only singularities are $s \geq (M_A + M_B)^2$. In what is now a classic, Martin /1966a/ made ingenious use of the positivity properties due to unitarity and the above analyticity properties to extend the aximoatic analyticity domain of $F(s,t)$. He proved that $F(s,t)$ has combined analyticity in s and in t in the domain

$$C = \Big\{(s,t)|t| < R,\ R > 0,\ s \in \text{cut plane with} \quad \text{cuts } s \geq (M_A + M_B)^2,\ u \geq (M_A + M_B)^2\Big\}. \tag{18}$$

/Sommer 1967/ then used the Lehmann ellipse to show that

$$R = Max_{s \geq (M_A + M_B)^2} \quad [2k^2(\cos\theta_0(s) - 1)], \tag{19}$$

with $\cos\theta_0$ given by (17). Using this and a result of Bessis and Glaser /1967/, we now know that

$$R = 4m_\pi^2, \tag{20}$$

for $\pi\pi$, KK, $K\bar{K}$, πK, πN and $\pi\Lambda$ scattering. Martin's reslut (18) means for the absorptive part $A_s(s,t)$ analyticity not only in $|t| < R$, but due to positivity, analyticity in the whole ellipse in t-plane with foci $t = 0$, $-4k^2$ and right extremity $t = R$. The s-independence of R plays the most crucial role in the proof of high energy bounds. Combining analyticity of $A_s(s,t)$ in this ellipse with analyticity in the large Lehmann ellipse, Martin proved fixed-t dispersion relations in a complex domain containing $|t| < 4m_\pi^2$ and extending from $Re\ t = -28m_\pi^2$ to $4m_\pi^2$ for $\pi\pi$ scattering, and from $Re\ t = -12.4m_\pi^2$ to $4m_\pi^2$ for πN scattering. Further, for pion-pion scattering, he combined analyticity in these two ellipses, the dispersion relation, and positivity properties to obtain the Lehmann-Martin ellipse of analyticity of $A_s(s,t)$. This ellipse has foci $t = 0$, $-4k^2$ and right extremity (in units $m_\pi = 1$)

$$\begin{aligned} t_0(s) &= 16 + \frac{64}{s-4}, \text{ for } 4 < s < 16, \\ t_0(s) &= 256/s, \text{ for } 16 < s < 32, \\ t_0(s) &= 4 + \frac{64}{s-16}, \text{ for } s > 32. \end{aligned} \tag{21}$$

This is bigger than the large Lehmann-ellipse for $s > 32$.

3.3 Crossing. The processes for which fixed-t dispersion relations have been proved do not include some interesting ones such as nucleon-nucleon scattering. The crossing property for an arbitrary process $A + B \to C + D$ proved by /Bros, Epstein and Glaser 1965/

$$F^{AB\to CD}(s,t) = F^{A\bar{D}\to C\bar{B}}(u,t), \tag{22}$$

for $t \leq 0$, and $|s|$ large enough, is therefore especially useful in the general proof of Pomeranchuk-like theorems. Bros, Epstein and Glaser showed that for any $t \leq 0$, $F^{AB\to CD}(s,t)$ is analytic in the cut-s plane except for possible singularities in a finite region and hence can be continued to the upper-lip of the left-hand cut where it equals the complex conjugate of the $A\bar{D} \to C\bar{B}$ physical amplitude. Recently Bros /1986/ has proved crossing for $2 \to 3$ particle processes.

4. High Energy Bounds.

4.1 Bounds on σ_{tot} /Martin 1963, 1966a/.

The fixed positive t dispersion relations obtained by Martin and the Lehmann-Martin ellipse of analyticity of $A_s(s,t)$ removed two important shortcomings of the Lehmann-ellipse. Convergence of the partial wave expansion of $A_s(s,t)$ within the large Lehmann-ellipse implies (using the large ℓ behaviour of $P_\ell(\cos\theta)$ for $\cos\theta > 1$)

$$\overline{\lim}_{\ell\to\infty}(Im\ a_\ell)^{1/\ell} \leq \frac{1}{Z_0+\sqrt{Z_0^2-1}} \underset{s\to\infty}{\sim} e^{\frac{-\text{const}}{s}}, \tag{23}$$

where we used $Z_0 \equiv 2\cos^2\theta_0 - 1 \sim 1 + \text{const}/s^2$ for $s \to \infty$. For smooth $Im\ a_\ell$, we have $Im\ a_\ell < exp(-\text{const}\ell/s)$, which means that the number of partial waves contributing effectively to the absorptive amplitude is $L_{max} \sim s$. This exceeds by far the semiclassical expectation $L_{max} \sim \sqrt{s}$ and moreover leads to a poor asymptotic upper bound $\sigma_{tot} < \text{const}.s\ln^2(s/s_0)$ /Greenberg and Low, 1961/. Martin's result of analyticity for $|t| < R$ immediately removes these two shortcomings. It implies

$$\overline{\lim}_{\ell\to\infty}(Im\ a_\ell)^{1/\ell} \leq e^{-\text{const}/\sqrt{s}}; \tag{24}$$

this gives roughly the expected L_{max} and leads to the Froissart-Martin bound on σ_{tot}. The proof consists in finding an upper bound on

$$\sigma_{tot} = \frac{4\pi}{k^2}\sum_{\ell=0}^{\infty}(2\ell+1)Im\ a_\ell(s) \tag{25}$$

given the positivity properties of $Im\ a_\ell$ and the polynomial boundedness of $A_s(s,t=R-\epsilon)$, $\epsilon > 0$,

$$A_s(s,t=R-\epsilon) = \frac{\sqrt{s}}{2k}\sum_{\ell=0}^{\infty}(2\ell+1)Im\ a_\ell(s)P_\ell\left(1+\frac{R-\epsilon}{2k^2}\right) < s^N. \tag{26}$$

Martin obtained

$$\sigma_{tot}(s) \underset{s\to\infty}{\leq} \frac{4\pi}{R}(1+\epsilon_1)(N-1)^2[\ln(s/s_0)]^2, \tag{27}$$

where ϵ_1 is an arbitrarily small positive number, and s_0 a constant. Martin similarly obtained using $|P_\ell(\cos\theta)| \leq 1$, for $-1 \leq \cos\theta \leq 1$, and $|a_\ell(s)| \leq \sqrt{Im\ a_\ell}$, the bound

$$|F(s,t)|_{t\leq 0} \; s \overset{<}{\to} \infty \frac{s}{4R}(1+\epsilon_1)N^2[\ln(s)/s_0)]^2. \tag{28}$$

Using crossing, on the left-hand cut also we have $|F(s,t)| < \text{const}|s|(\ln|s|)^2$; the Phragmen-Lindelöf theorem then gives the same bound in the whole cut-plane. Hence we need at most two subtractions for $-t_1 \leq t \leq 0$; by the Jin-Martin /1964/ theorem we therefore need at most two subtractions also for $|t| < R$. Hence $N = 2$, and using $R = 4m_\pi^2$ for $\pi\pi$, KK, $K\bar{K}$, πK, πN abd $\pi\Lambda$ scattering, we have for these processes the Froissart-Martin bound,

$$\sigma_{tot}(s) \; s \overset{<}{\to} \infty \frac{\pi}{m_\pi^2}(1+\epsilon_1)[\ln(s/s_0)]^2 \tag{29}$$

It is impresive that the coefficient (π/m_π^2) following from general principles has the reasonable value of about 60 millibarns. The unknown coefficient s_0 is a problem if we want a direct experimental test. This difficulty was solved by Yndurain /1970/, Common /1979/, and Roy /1972/ who obtained upper bounds on energy averages of σ_{tot} using the extra input of the experimental D-wave $\pi\pi$ scattering lengths and found results within a factor 4 of the experimental values. Similar bounds without this extra input (and hence more rigorous though numerically weaker) follow from the absolute bounds on pion-pion scattering /Martin 1965a; Lukaszuk and Martin 1967/.

It is interesting that the general principles also imply a lower bound on σ_{tot}, albeit poor. Improving the result $\sigma_{tot} \geq \sigma_{el} \geq \text{const} s^{-6}[\ln(s/s_0)]^{-2}$ of Jin and Martin /1964/, Cornille /1971/ proved that

$$\sigma_{tot}(s) s \overset{\geq}{\to} \infty \frac{\text{const}}{s^6} \tag{30}$$

for at least one sequence of $s \to \infty$.

4.2 Bounds on Differential Cross sections and Total Cross section Differences.

/Eden 1966, Kinoshita 1966, Bessis 1966, Martin 1963b, Logunov and Van Hieu 1968/ observed that the Froissart-Martin bounds on differential cross sections could be improved if the integrated (elastic or inelastic) cross sections were known. Determination of unknown constants in these bounds leads to the final result /Singh and Roy, 1970, Roy and Singh 1970/,

$$\left(\frac{d\sigma}{d\Omega}\right)^{AB\to CD} \underset{s\to\infty}{\lesssim} \left\{(L+1)^2(P_L(\cos\theta))^2 + \sin^2\theta(P'_L(\cos\theta))^2\right\}, \quad (31)$$

where

$$L = \frac{1}{2}\sqrt{s/R}\ \ln\left(\frac{s}{s_0^2\sigma^{AB\to CD}}\right), \quad (32)$$

$$\sigma^{AB\to CD} = \int d\Omega (d\sigma/d\Omega)^{AB\to CD}, \quad (33)$$

and R is given by (20). [The usual Froissart-Martin bound can be derived from this by replacing $\sigma^{AB\to CD}$ by its upper bound (29)]. For forward scattering this implies

$$\left(\frac{d\sigma}{d\Omega}\right)^{AB\to CD}_{\theta=0} \underset{s\to\infty}{\lesssim} \frac{s}{16\pi R}\sigma^{AB\to CD}\left[\ln\left(\frac{s}{s_0^2\sigma^{AB\to CD}}\right)\right]^2 \quad (34)$$

For $\pi^- p \to \pi^0 n$ scattering, <u>if isotopic spin invariance holds</u>, using $|Im\ F(s,0)| \leq |F(s,0)|$, Eq. (34) gives the very useful bound

$$|\sigma_{tot}^{\pi^- p}(s) - \sigma_{tot}^{\pi^+ p}(s)| \underset{s\to\infty}{\lesssim} \frac{\sqrt{2\pi}}{m_\pi}\sqrt{\sigma^{\pi^- p\to\pi^0 n}(s)}\ \ln\left(\frac{s}{s_0^2\sigma^{\pi^- p\to\pi^0 n}}\right). \quad (35)$$

If, in addition to (34) forward dispersion relations and iso-spin invariance are used we obtain /Roy and Singh 1970/

$$|\lim_{s\to\infty}[\sigma_{tot}^{\pi^- p} - \sigma_{tot}^{\pi^+ p}]| \leq \frac{\pi^{3/2}}{m_\pi\sqrt{2}}\lim_{s\to\infty}\sqrt{\sigma^{\pi^- p\to\pi^0 n}} \quad (36)$$

provided the above limits exist. At Serpukhov /Apel et al 1979/ measurement upto 40 GeV/c pion momentum yields

$$\sigma^{\pi^- p\to\pi^0 n} = (122\pm 8)(s/10)^{-1.23\pm 0.02}\times 10^{-30}cm^2, \quad (37)$$

where s is in GeV^2 units. At Fermilab /A.S. Carroll et al 1976/ measurement upto 240 GeV/c pion-momentum yields

$$\sigma_{tot}^{\pi^- p} - \sigma_{tot}^{\pi^+ p} = As^{\alpha-1}, \quad \alpha = 0.55\pm .03. \quad (38)$$

The experimental fits (37), (38) if extrapolated to $s\to\infty$ violate the bound (35) by four standard deviations. Martin suggested that higher energy measurements could now be made at Fermilab to probe this question further.

4.3 Bounds on Absorptive Parts and Absorptive Differential cross sections.

Bounds valid at any energy (not just asymptotic energies) can be obtained from the positivity properties of the absorptive part and absorptive differential cross sections. The fore-runner of these is the /Mac Dowell-Martin 1964/ bound on the absorptive elastic diffraction peak width for spinless case

$$\frac{d}{dt}\ln\left(\frac{d\sigma_A}{dt}\right)\Big|_{t=0} \geq \frac{2}{9}\left[\frac{\sigma_{tot}^2}{4\pi\sigma_{el}} - \frac{1}{k^2}\right]. \tag{39}$$

In the high energy limit, for arbitrary spin, /Auberson, Martin and Mennessier 1977/ found the above result with an extra factor 0.9965 on the right-hand side. For πN scattering, at any energy, /Mennessier, Roy and Singh 1979/ proved that

$$\frac{d}{dt}\ln\frac{d\sigma_A}{dt}\Big|_{t=0} > \frac{\sigma_{tot}^2}{36\pi\sigma_{el}}\left[1+\left\{1+\frac{9\pi\sigma_{el}}{(2k\sigma_{tot})^2}\right\}^{1/2}\right] - \frac{1}{4k^2}. \tag{40}$$

Roy /1979/ proved that for $-1 \leq \cos\theta \leq 1$,

$$[f(\cos\theta)/f(1)]^2 \leq \frac{1}{3} + \frac{2}{3}f\left(\frac{1}{2}(3\cos^2\theta - 1)\right)/f(1), \tag{41}$$

where $f(\cos\theta) = Im\ F(s,\cos\theta)$ or $d^n Im\ F(s,\cos\theta)/d(\cos\theta)^n$ for spinless elastic scattering or the analogous quantities with $Im\ F \to d\sigma_A/dt$ for arbitrary spin elastic scattering. Singh and Roy /1970a,b/ proved that for elastic spinless scattering,

$$\left|\frac{Im\ F(s,t)}{Im\ F(s,0)}\right|_{s\to\infty}^{\leq} \left[1 - \frac{r}{9} + \frac{3}{8}\left(\frac{r}{9}\right)^2 - \frac{21}{320}\left(\frac{r}{9}\right)^3 + \cdots\right], \tag{42}$$

if

$$0 \leq r \equiv (-t)\sigma_{tot}^2/(4\pi\sigma_{el}) \leq 2.5. \tag{43}$$

Only the first three terms on the right-hand side of (42) are necessary for an accuracy of 0.5 per cent. Generalizations in several directions are summarized by /Valin 1989/. The bounds (39) - (42) are numerically very close to and sometimes apparently violated by experimental data. Hence they are useful as unitarity tests on data.

The asymptotic bound

$$b(s) \equiv \frac{d}{dt}\ln\left(\frac{d\sigma_A}{dt}(s,t)\right)\Bigg|_{t=0} \ \underset{s\to\infty}{\leq}\ \frac{1}{4R}\left[\ln\left(\frac{s^2}{\frac{d\sigma^A}{dt}(s,0)}\right)\right]^2 \equiv b_{Max}(s), \tag{44}$$

derived by /Singh 1971/ for spinless case and by /Auberson and Roy 1976/ for general spins is useful in discussion of scaling properties.

5. <u>Saturation of High Energy Bounds and Scaling.</u>

Auberson, Kinoshita and Martin /1971/ proved the following scaling theorem for the spinless case. If $\sigma_{tot}^{AB} \sim \text{const.}(\ell n(s/s_0))^2$ for $s \to \infty$, then every sequence of $s \to \infty$ must contain a subsequence such that

$$\lim_{s \to +\infty, \tau - \text{fixed}} \frac{F^{AB \to CD}(s, t = -\tau[\ln(s/s_0)]^{-2})}{F^{AB \to CD}(s, 0)} = f(\tau), \tag{45}$$

where $f(\lambda)$ is an entire function of τ of order half.

Auberson and Roy /1976/ proved the following theorem for the absorptive contribution to the elastic unpolarized differential crosssection for arbitrary spin processes. If the bound (44) is qualitatively saturated, i.e., if $b(s)/b_{Max}(s) \geq b_0 \neq 0$, for $s \to \infty$, then, every squence of $s \to \infty$ must contain a subsequence such that

$$\lim_{s \to \infty, \tau \text{ fixed}} \left[\frac{d\sigma_A}{dt}(s, t = -\frac{\tau}{b(s)}) / \frac{d\sigma_A}{dt}(s, 0) \right] = f(\tau), \tag{46}$$

where $f(\tau)$ is an entire function of order half obeying $f(0) = 1$, $f'(0) = -1$, and

$$f(\tau) = \int_{\lambda=0}^{2/\sqrt{b_0}} d\mu(\lambda) J_0(\lambda\sqrt{\tau}), \tag{47}$$

where $d\mu(\lambda)$ is a positive measure obeying

$$\int_{\lambda=0}^{2/\sqrt{b_0}} d\mu(\lambda) = 1, \qquad \int_{\lambda=0}^{2/\sqrt{b_0}} d\mu(\lambda)\, \lambda^2 = 4. \tag{48}$$

The second theorem above is of more general applicability because whenever $\sigma_{tot} \sim (\ln(s/s_0))^2$, the bound (39) shows that $b(s)$ also $\sim [\ln(s/s_0)]^2$ and hence the bound (44) is then qualitatively saturated. On the otherhand the first theorem (45) also covers the case of dominant real parts.

6. Pomeranchuk-like Theorems.

Martin /1965a/ proved one of the early versions of the Pomeranchuk theorem. If

$$\lim_{s\to\infty}\left[\sigma_{tot}^{AB}-\sigma_{tot}^{\bar{A}B}\right] \tag{49}$$

exists, the limit being finite or infinite, and if

$$\lim_{s\to\infty}\frac{F(s,t=0)}{s\ln(s/s_0)}=0, \tag{50}$$

then the limit (49) must be zero.

Another version is due to /Eden 1966, Kinoshita 1966/, and /Grunberg and Truong 1973, 1974/. If σ_{tot}^{AB} or $\sigma_{tot}^{\bar{A}B}\to\infty$ for $s\to\infty$, then the ratio $\sigma_{tot}^{AB}/\sigma_{tot}^{\bar{A}B}$ approaches unity if the ratio has a limit (otherwise unity is among the limiting values of this ratio). Experimentally $\sigma_{tot}^{\bar{p}p}/\sigma_{tot}^{pp}\approx 1.065$ at $E_{lab}=$ 200 GeV.

If isotopic spin invariance is valid, then the result (36) implies that if $\lim\sigma^{\pi^-p\to\pi^0n}=0$, then $\lim[\sigma_{tot}^{\pi^-p}-\sigma_{tot}^{\pi^+p}]=0$.

More recently the following theorem has been derived by rather impressive use of positivity properties and analyticity by Cornille and Martin /1972a,b,c;1974/. Inside the diffraction peak, the ratio $d\sigma/dt^{AB\to AB}(s,t(s))$ $/(d\sigma/dt)^{\bar{A}B\to\bar{A}B}(s,t(s))$ approaches unity as $s\to\infty$, for any sufficiently smooth $t(s)$ such that $(d\sigma/dt)(s,0)/(d\sigma/dt)(s,t(s))$ stays finite as $s\to\infty$. If the limit does not exist, unity naturally belongs to the set of limiting values. It follows that the ratio of slope parameters also $\to 1$. Experimentally /Favart et al 1981/, at $|t|<.03$ GeV^2 and $E_{lab}=1,500$ GeV we have $b^{\bar{p}p}/b^{pp}=1.015\pm 0.04$, in very good agreement with the theorem.

References

Apel, W.D. et al /1979/ Nucl. Phys. B154, 189.

Auberson, G., T. Kinoshita and, A. Martin /1971/ Phys. Rev. D3, 3185.

Auberson, G., and S.M. Roy /1976/ Nucl. Phys. B117, 322.

Auberson, G., A. Martin and G. Mennessier /1977/ Phys. Lett. 67B, 75.

Besis, J.D. /1966/ Nuovo Cim. 45A, 974.

Bessis, J.D. and V. Glaser /1967/ Nuovo. Cim. 58, 568.

Bogoliubov, N.N., B.V. Medvedev and M.K. Polivanov /1958/ Voprossy Teorii Dispersionykh Sootnoshenii, Moscow.

Bremermann, H.J., R. Oehme and J.G. Taylor /1958/ Phys. Rev. 109, 2178.

Bros, J., H. Epstein and V. Glaser /1965/ Comm. Math. Phys. 1, 240.

Bros, J. /1986/ in "Essays in Mathematical Physics in Memoriam V. Glaser" ed. A. Martin and M. Jacob, Physics Reports 134, 325.

Carroll, A.S. et al /1976/ Phys. Lett. 61B, 303.

Cornille, H. /1971/ Nuovo. Cim. 4A, 549.

Cornille, H., A. Martin /1972a/: Phys. Lett. 40B, 671.

Cornille, H., A. Martin /1972b/: Nucl. Phys. B48, 104.

Cornille, H., A. Martin /1972c/: Nucl. Phys. B49, 413.

Cornille, H., A. Martin /1974/: Nucl. Phys. B77, 141.

Cornille, H., A. Martin /1976/ Nucl. Phys. B115, 163.

Eden, R.J. /1966/ Phys. Rev. Lett. 16, 39.

Eden, R.J. /1971/ Rev. Mod. Phys. 43, 15.

Epstein, H., V. Glaser and A. Martin /1969/ Comm. Math. Phys. 13, 257.

Favart, D. et al /1981/ Phys. Rev. Lett. 47, 1191.

Fischer, J. /1981/ Phys. Reports 76, 157.

Froissart, M. /1961/: Phys. Rev. 123, 1053.

Greenberg, O.W. and F.E. Low /1961/ Phys. Rev. 124, 2047.

Grunberg, G., and T.N. Truong /1973/ Phys. Rev. Lett. 31, 63

Grunberg, G., and T.N. Truong /1974/ Phys. Rev. D9, 2874.

Greenberg, O.W. and F.E. Low /1961/ Phys. Rev. 124, 2047.

Jin, Y.S., A. Martin /1964/ Phys. Rev.: 135, B 1369.

Kinoshita, T., /1966/ "Perspectives in Modern Physics", Ed. R.E. Marshak (Wiley, New York 1966), p.211.

Lehmann, H. /1958/: Nuovo Cim. 10, 579.

Lehmann, H. /1959/: Fort. d. Physik VI, 159.

Logunov, A.A. and N. Van Hieu /1968/ Proc. topical conference on high energy collisions of hadrons, Vol. II, p.74, CERN, Geneva.

Lukaszuk, L., A. Martin /1967/: Nuovo Cim. 52A, 122.

Martin, A. /1963a/: Phys. Rev. 129, 1432.

Martin, A. /1963b/: Nuovo Cim. 29, 993.

Martin, A. /1965/: "High Energy Physics and Elementary Particles", I.A.E.A., Vienna,. p.155.

Martin, A. /1965a/: Nuovo Cim. 39, 704.

Martin, A. /1966a/: Nuovo Cim. 42, 930.

Martin, A. /1966b/: Nuovo Cim. 44, 1219.

Martin, A. /1969/: "Scattering theory: Unitarity, Analyticity and Crossing" (Springer-Verlag, Berlin, Heidelberg, New York 1969).

Martin, A., F. Cheung /1970/ "Analyticity Properties and Bounds of the Scattering Amplitudes", Gordon and Breach, New York 1970.

Mahoux, G. /1976/ Phys. Lett. 65B, 139.

Mennessier, G., S.M. Roy and V. Singh /1979/ Nuovo Cim. 50A, 443.

Roy, S.M. and V. Singh /1970/ Phys. Lett. 32B, 50.

Roy, S.M. /1972/ Physics Reports 5C, 125.

Roy, S.M. /1977/ Phys. Lett. 70B, 213.

Roy, S.M. /1979/ Phys. Rev. Lett. 43, 19.

Singh, V. and S.M. Roy /1970/ Ann. Phys. 57, 461.

Singh, V. and S.M. Roy /1970a/ Phys. Rev. Lett. 24, 28.

Singh, V. and S.M. Roy /1970b/ Phys. rev. 1D, 2638.

Singh, V. /1971/ Phys. Rev. Lett. 26, 530.

Singh, V. /1971/ Fields and Quanta 1, 151.

Sommer, G. /1967/ Nuovo. Cim. 48A, 92.

Symanzik, K. /1957/ Phys. Rev. 100, 743.

Valin, P. /1989/ McGill University Reprint

Complex Angular Momentum Analysis in Axiomatic Quantum Field Theory

J. Bros[1] *and G.A. Viano*[2]

[1]Service de Physique Théorique,
F-91191 Gif-sur-Yvette Cedex, France
[2]Istituto Nazionale di Fisica Nucleare (I.N.F.N.),
Sezione di Genova, Dipartimento di Fisica dell'Universita,
Genova, Italy

1. INTRODUCTION AND DESCRIPTION OF THE RESULTS

In this paper, we wish to present a missing item in the list of analyticity properties rigorously derived from the axioms of Quantum Field Theory (Q.F.T.), part of which served as a basis for the numerous results on relativistic scattering amplitudes, due to the ingenuity of André Martin.

This missing item, which concerns the introduction of analyticity with respect to *complex angular momentum variables* for appropriate Fourier-Laplace type transforms of the Q.F.T. Green functions, will be fully developed elsewhere[B.V.].

In the present survey, we will try to show that Regge's philosophy concerning analyticity and singularities in the complex angular momentum plane (set on a firm mathematical basis in the framework of non-relativistic potential theory[Re]) is not only "consistent with", but "conceptually implied by" the axiomatic framework of Quantum Field Theory (see e.g. [S,Wi]), provided one takes as primitive objects of study (as in many problems of Q.F.T.) the off-shell "euclidean amplitudes" instead of taking abruptly the physical (on-shell) scattering amplitudes. In particular, we shall establish here a field-theoretical off-shell version of the famous Froissart-Gribov representation[F][G] of the partial waves, which was discovered by these authors in 1961 in the analytic $S-$Matrix approach of Particle Physics. In this connection, it is a pleasure for us to emphasize in these lines that the Froissart-Gribov representation was also derived independently, although slightly later, by André Martin in a clear and beautiful paper[Ma-1] which contained further interesting results, such as the extension of unitarity to complex angular momenta. (Related contemporary, but unpublished works of R. Omnes and E.J. Squires must also be mentioned).

This first section is devoted to a short description of our results; some complements on the proofs and on the mathematical methods involved will be given in section 2 and in the Appendix; finally, after a comparative historical survey of related works outside the framework of axiomatic Q.F.T., we indicate in section 3 a possible scheme for further investigations.

Let $F(k_1,k_2,k_1',k_2') \equiv F[(\zeta_1,\zeta_2),(\zeta_1',\zeta_2');t,\cos\Theta_t]$ be the four-point function of a scalar Q.F.T. in $(d+1)$-dimensional space-time ($d \geq 2$), considered through the kinematics of a *distinguished two-particle channel* with squared energy-momentum $t=(k_1+k_2)^2=(k_1'+k_2')^2$ and corresponding scattering angle Θ_t; k_i, k_i' $(i=1,2)$ denote complex $(d+1)$-momenta such that $k_1+k_2=k_1'+k_2'$: $\zeta_i=k_i^2$, $\zeta_i'=k_i'^2$ are the corresponding squared-mass variables (if $k=\left(k^{(0)},\vec{k}\right)\in \mathbf{C}^{d+1}$, we put: $k^2={k^{(0)}}^2-\vec{k}^2$) and we use the notations $\zeta=(\zeta_1,\zeta_2)$, $\zeta'=(\zeta_1',\zeta_2')$.

Our purpose is to introduce and study an integral transform $\tilde{F}(\zeta,\zeta';t,\lambda)$ of $F[\zeta,\zeta';t,\cos\Theta_t]$ which interpolates analytically in an appropriate way the set of (off-shell) t-channel partial-waves:

$$f_\ell^{(d)}(\zeta,\zeta',t)=\int_{-1}^{+1}P_\ell^{(d)}(\cos\Theta_t)\,F[\zeta,\zeta';t,\cos\Theta_t]\,[\sin\Theta_t]^{d-3}\,\mathrm{d}\cos\Theta_t \qquad (1)$$

in a domain of the complex λ-plane. In the latter, the functions $P_\ell^{(d)}$ are generalized Legendre polynomials (which reduce to $\cos\ell\Theta_t$ for $d=2$, to the Legendre polynomials P_ℓ for $d=3$ and to the Gegenbauer polynomials for $d=4$).

In this study as in the axiomatic scheme for proving analyticity properties in complex momentum space, two parts may be distinguished.

1.1 The transform $\tilde{F}$ of the four-point function F in the "linear program"

This first part makes use of a basic analyticity property of F, implied by the standard axioms of locality, spectrum and Lorentz invariance (i.e. pertaining to the so-called "linear program" of axiomatic Q.F.T.): this property is the fact that $F[\zeta,\zeta';t,\cos\Theta_t]$ is analytic in a cut-plane of the variable $\cos\Theta_t$ (with cuts contained in the half-lines $(-\infty,-1]$ and $[+1,+\infty)$), for all (ζ,ζ',t) belonging to a certain off-shell region $\Delta^{(e)}$. (see §2-2 for some account on the derivation of the latter).

On the other hand, this study also relies in a crucial way on the validity of a power-type majorization of the form: $|F[\zeta,\zeta';t,\cos\Theta_t]|\leq C_{\zeta,\zeta',t}\,|\cos\Theta_t|^N$ in the cut-plane considered; this majorization is itself a consequence of the "temperateness axiom" of Q.F.T.

The basic result which can be derived from the latter properties is the existence (for any dimension $d \geq 2$) of an appropriate transform $\tilde{F}(\zeta,\zeta';t,\lambda)$ of F, analytic in the corresponding half-plane $\Pi_N = \{\lambda; \mathrm{Re}\ \lambda > N\}$ and defined for all (ζ, ζ', t) in the region $\Delta^{(e)}$.

Let F_+ and F_- be the jumps of the function iF across the respective cuts σ_+ and σ_- contained in the half-lines $[1, +\infty)$ and $(-\infty, -1]$ of the $\cos\Theta_t$-plane. These cuts represent respectively (in the off-shell situation $(\zeta, \zeta', t) \in \Delta^{(e)}$ considered) the crossed channel s-cut $(s = (k_1 - k_1')^2 \geq s_0)$ and u-cut $(u = (k_1 - k_2') \geq u_0)$ and F_+, F_- are the corresponding *absorptive parts* of F.

We can then state in a more specific way the basic result of this first part as follows:

Property I: Let $\Delta^{(e)} = \{(\zeta, \zeta', t);\ t \leq 0,\ \zeta \in \Delta_t, \zeta' \in \Delta_t\}$, where Δ_t is the parabolic region $\{\zeta = (\zeta_1, \zeta_2);\ \zeta_i \leq 0, i = 1, 2; \Lambda(\zeta_1, \zeta_2, t) \equiv (\zeta_1 - \zeta_2)^2 - 2(\zeta_1 + \zeta_2)t + t^2 \leq 0\}$ Then, for all (ζ, ζ', t) in $\Delta^{(e)}$, the subset of partial-waves $\left\{f_\ell^{(d)}; \ell \geq N\right\}$ is interpolated in the half-plane Π_N by an analytic function $\tilde{F}(\zeta, \zeta'; t, \lambda)$ which satisfies the following properties

a) $$\tilde{F} = \tilde{F}_+ + e^{i\pi\lambda}\, \tilde{F}_-\ , \tag{2}$$

where

$$\tilde{F}_\pm(\zeta, \zeta'; t, \lambda) = \pm \int_{\sigma_\pm} Q_\lambda^{(d)}(\cos\Theta_t)\, F_\pm[\zeta, \zeta'; t, \cos\Theta_t]\, (\sin\Theta_t)^{d-3}\ \mathrm{d}\cos\Theta_t \tag{3}$$

in the latter, the functions $Q_\lambda^{(d)}$ are generalized second-kind Legendre functions.

b) $\tilde{F}_+$ and $\tilde{F}_-$ are analytic for λ in Π_N and uniformly bounded (up to temperate factors) by $e^{-\mu_\pm(\mathrm{Re}\ \lambda)}$, where μ_+, μ_- are positive numbers determined respectively by the threshold s_0 and u_0 of the s- and u-cuts.

An alternative way of stating property I is obtained by introducing the symmetric and antisymmetric combinations: $\tilde{F}^{(s)} = \tilde{F}_+ + \tilde{F}_-$ and $\tilde{F}^{(a)} = \tilde{F}_+ - \tilde{F}_-$, namely.

Second form of Property I:

The transform $\tilde{F}$ of F has the following structure:

$$\tilde{F} = e^{i\pi\frac{\lambda}{2}} \left[\cos\frac{\pi\lambda}{2}\tilde{F}^{(s)} - i\sin\frac{\pi\lambda}{2}\tilde{F}^{(a)}\right], \tag{4}$$

where $\tilde{F}^{(s)}$ and $\tilde{F}^{(a)}$ are Carlsonian interpolations in the half-plane Π_N for the respective sets of *even* and *odd* partial-waves, namely one has:

$$\tilde{F}^{(s)}(\zeta, \zeta'; t, \lambda)|_{\lambda = 2\ell} = f_{2\ell}(\zeta, \zeta', t) \tag{5}$$

and

$$\tilde{F}^{(a)}\left(\zeta,\zeta';t,\lambda\right)|_{\lambda=2\ell+1} = f_{2\ell+1}\left(\zeta,\zeta',t\right) \tag{5'}$$

The previous statements call for two comments:

i) The region $\Delta^{(e)}$ for which Property I has been stated corresponds to the image (in (ζ,ζ',t)-space of all euclidean momentum configurations (i.e. of the sets $(k_i, k'_i;\ i = 1,2)$ such that $k_i = \left(i\ q_i^{(0)}, \vec{p}_i\right)$, $k'_i = \left(i\ q_i'^{(0)}, \vec{p}_i{}'\right)$). The problem of extending the validity of Property I to a larger region Δ of (ζ,ζ',t)-space is not treated here; it corresponds to an exploration of the largest domain for which (off-shell or on-shell...) dispersion relations in the cos Θ_t-plane can be derived in the general framework of Q.F.T. (see our comments in section 3).

ii) The occurrence of second-kind Legendre functions and of the absorptive parts of F under the integration sign of formula (3) will of course invite the reader to consider Property I as an "off-shell counterpart" (in dimension d) of the Froissart-Gribov representation[F][G] of the physical partial-waves. As a matter of fact, making such a parallel is both pertinent and incomplete... It is pertinent since the F.G. representation was based on a mathematical structure which is identical with the present one, namely the analyticity and power-boundedness of the scattering amplitude, in a cut-plane of the complex variable cos Θ_t. In that case, however, this structure was *not* implemented by axiomatic Q.F.T., but it was based on the requirement that the scattering amplitude should satisfy the Mandelstam representation[M].

On the other hand, the parallel is incomplete for the following reason: the Q.F.T. framework leads one to an insight into the mathematical content of F.G.-type representations which goes beyond the pure technical use of suitable properties of the special functions $P_\ell^{(d)}$ and $Q_\lambda^{(d)}$. In fact, as it is explained in section 2 (§2.1), this mathematical content can be considered from a more comprehensive geometrical viewpoint, inspired by the previous work of Faraut and Viano[F,V], in which the Legendre functions $P_l^{(d)}$ and $Q_\lambda^{(d)}$ are introduced through appropriate integral representations. This viewpoint presents the following advantages:

a) it exhibits clearly (in any dimension d) the group-theoretical foundation of the property considered in terms of generalized Fourier-Laplace transformations;

b) it involves a geometrical situation which is built-in in the complex momentum space scenario of Q.F.T. (see §2.2);

c) it is essential for obtaining the second type of structural property of $\tilde{F}$ provided by Q.F.T., which we shall now describe.

1.2 Meromorphic continuation of $\tilde{F}$ in the "non-linear program"

This second part concerns a procedure which potentially generates a meromorphic continuation of $\tilde{F}(\zeta,\zeta';t,\lambda)$, with the occurrence of poles in the joint variables (λ,t), under appropriate assumptions to be added to the axioms of local fields. In the enriched version of axiomatic Q.F.T. relying on an additional postulate of Asymptotic Completeness, the meromorphic continuation of F with possible poles in the variable t (in a ramified neighbourhood of the two-particle region) results[B,BL,BI] from the introduction of a general "two-particle irreducible kernel" $G(k_1,k_2;k_1',k_2') \equiv G[\zeta,\zeta';t,\cos\Theta_t]$ related to F by a Bethe-Salpeter-type equation of the form:

$$F = G + F \circ_t G\,; \tag{6}$$

in the latter, $\circ_t$ denotes a (possibly regularized) two-line Feynman-type convolution with respect to the t channel which is initially integrated on the euclidean subspace. $\circ_t$ can be decomposed into an integration operation $\tilde{\circ}_t$ over the domain Δ_t, with respect to the mass variables $\zeta'' = (\zeta_1'',\zeta_2'')$ carried by the intermediate two lines and a convolution-product (in the sense of $SO(d)$) on the sphere S_{d-1}. In view of rotational invariance properties of F and G, Eq.(6) then yields the following system of equations for the partial waves f_ℓ of F:

$$\forall \ell \in \mathbb{N}, \qquad f_\ell(\bullet;t) = g_\ell(\bullet;t) + (f_\ell \tilde{\circ}_t g_\ell)(\bullet;t) \tag{7}$$

in the latter, the dots stand for (ζ,ζ'), the functions $g_\ell(\zeta,\zeta';t)$ are the "partial-waves" of G defined by equations similar to Eqs.(1), and the operation $\tilde{\circ}_t$ is specified as follows:

$$(a\ \tilde{\circ}_t\ b)(\zeta,\zeta';t) = \int_{\Delta_t} a(\zeta,\zeta'';t)\, b(\zeta'',\zeta';t) \left[\frac{\lambda(\zeta_1'',\zeta_2'',t)}{4t}\right]^{\frac{d-3}{2}} \frac{d\zeta_1'' d\zeta_2''}{4\sqrt{-t}} \tag{8}$$

The analyticity domain of the functions f_ℓ (which can be deduced from that of F) contains $\Delta^{(e)}$, and is in fact much larger since it also includes a ramified neighbourhood of the t-channel two-particle region on the mass-shell (see our comments in section 3).

In the framework of this "non-linear program" of axiomatic Q.F.T., one is then led to investigate the consequences of the B.S. equation (6) for the analytic structure of the transform $\tilde{F}(\zeta,\zeta';t,\lambda)$ of F, introduced above.

Since G admits the same analyticity domain as F (except for possible poles in t)[B], it is reasonable to assume that for all $(\zeta,\zeta';t)$ in $\Delta^{(e)}$, G also satisfies a bound of the form $|G(\zeta,\zeta';t,\cos\Theta_t)| \le C'_{\zeta\zeta' t}\,|\cos\Theta_t|^{N'}$ where N' is not necessarily equal to N; Property I then applies to G and allows one to introduce its transform $\tilde{G}(\zeta,\zeta';t,\lambda)$ in terms of functions $\tilde{G}_+$, $\tilde{G}_-$, or $\tilde{G}^{(s)}$, $\tilde{G}^{(a)}$ associated with the $s-$ and $u-$channel absorptive parts G_+, G_- of G via formulae similar to Eqs.(2),...,(5'). Let then $\bar{N} = \max(N,N')$ and $\underline{N} = \min(N,N')$. The basic result in this second part is the "partial diagonalization" (in the specific sense explained below) of the Bethe-Salpeter equation (6) in the complex variable λ.

Property II: The following relations hold, for (ζ,ζ',t) in $\Delta^{(\epsilon)}$ and λ in $\Pi_{\bar{N}}$:

$$\tilde{F}^{(s)}(\bullet;t,\lambda) = \tilde{G}^{(s)}(\bullet;t,\lambda) + \left(\tilde{F}^{(s)}\tilde{\circ}_t\tilde{G}^{(s)}\right)(\bullet;t,\lambda) \tag{9}$$

$$\tilde{F}^{(a)}(\bullet;t,\lambda) = \tilde{G}^{(a)}(\bullet;t,\lambda) + \left(\tilde{F}^{(a)}\tilde{\circ}_t\tilde{G}^{(a)}\right)(\bullet;t,\lambda) \tag{10}$$

where the dots stand for (ζ,ζ') and $\tilde{\circ}_t$ denotes the integration operation defined (for each λ) by Eq.(8). Moreover, the restrictions of Eqs.(9) and (10) to the respective sets of even and odd integers $\lambda = \ell \ge \bar{N}$ yield the corresponding equations (7) for the partial waves f_ℓ.

Once Eqs.(9) and (10) are established, the latter statement is of course a direct consequence of Property I.

The proof of Eqs.(9) and (10) relies on the fact (established in the axiomatic framework in [B]) that the (t-channel) Bethe-Salpeter equation (6) implies the following coupled equations for the absorptive parts of F *in the s- and u-channels:*

$$F_+ = G_+ + F_+ \diamond_t G_+ + F_- \diamond_t G_-, \tag{11}$$

$$F_- = G_- + F_- \diamond_t G_+ + F_+ \diamond_t G_-, \tag{12}$$

in the latter, $\diamond_t$ denotes a Feynman-type convolution, whose integration cycle is a *compact* (double-cone-shaped) region, determined by the support properties (due to spectrum) of the absorptive parts involved. As a matter of fact (see [F.V.]), $\diamond_t$ can be decomposed into the integration operation $\tilde{\circ}_t$ in the mass variables (see Eq.(8)) and an integral in (hyperbolic) angular variables, interpretable as a certain convolution-product on an appropriate orbit of the group $SO(1,d-1)$, namely a one-sheeted hyperboloid (see below in §2.2). This convolution product, introduced and studied by Faraut and Viano[F.V.] enjoys the property of being changed into an ordinary product in the "Fourier-type variable" λ (see below in §2.1). Eqs.(11), (12) are thus shown to be equivalent to corresponding "partially diagonalized" equations for the quantities $\tilde{F}_\pm$ and $\tilde{G}_\pm$, in which each term $F_\varepsilon \diamond_t G_{\varepsilon'}$ (with $\varepsilon,\varepsilon' = +$ or $-$)

is replaced by the corresponding term $\left(\tilde{F}_\epsilon \tilde{\circ}_t \tilde{G}_{\epsilon'}\right)(\bullet; t, \lambda)$; finally, this system of equations takes the more tractable (decoupled) form of Eqs.(9), (10) in terms of the functions $\tilde{F}^{(s)}$, $\tilde{G}^{(s)}$ and $\tilde{F}^{(a)}$, $\tilde{G}^{(a)}$.

The Fredholm formulae can be applied to Eqs.(9), (10) not only in $\Pi_{\bar{N}}$, but in $\Pi_{\underline{N}}$ (namely, for computing either $\tilde{F}$ in terms of $\tilde{G}$ or $\tilde{G}$ in terms of $\tilde{F}$ according to whether $\underline{N} = N'$ or N). In the more interesting case when $\underline{N} = N' < N$, Property II admits therefore the following

Corollary: The transform $\tilde{F}(\zeta, \zeta'; t, \lambda)$ of F, primitively defined and analytic in Π_N (for $(\zeta, \zeta'; t) \in \Delta^{(e)}$) admits a meromorphic continuation in the strip $\Pi_{N'} \setminus \Pi_N = \{\lambda \in \mathbb{C}; N' < \text{Re}\,\lambda \leq N\}$, given by formula (4), with the following specification for the functions $\tilde{F}^{(s)}$ and $\tilde{F}^{(a)}$

$$\tilde{F}^{(s)}(\zeta, \zeta'; t, \lambda) = \frac{A^{(s)}(\zeta, \zeta'; t, \lambda)}{D^{(s)}(t, \lambda)}, \tag{13}$$

$$\tilde{F}^{(a)}(\zeta, \zeta'; t, \lambda) = \frac{A^{(a)}(\zeta, \zeta'; t, \lambda)}{D^{(a)}(t, \lambda)}; \tag{14}$$

in these formulae, the functions $A^{(s)}$ and $D^{(s)}$ (resp. $A^{(a)}$ and $D^{(a)}$) are analytic for λ in $\Pi_{N'}$, and computable in terms of $\tilde{G}^{(s)}$ (resp. $\tilde{G}^{(a)}$).

2. COMPLEMENTS

2.1 Mathematical results

In its bare form, the mathematical content of Froissart-Gribov-type representations is a theorem of complex analysis in one variable. It claims that for each integer d $(d \geq 2)$ there is a bijection between classes of functions $F[\cos\theta]$ analytic and bounded by $C^t[\cos\theta]^N$ in a cut-plane, with cuts σ_+, σ_- embedded respectively in $[+1, +\infty)$ and $(-\infty, -1]$, and corresponding classes of functions $\tilde{F}^{(d)}(\lambda) = \tilde{F}_+^{(d)}(\lambda) + e^{i\pi\lambda}\tilde{F}_-^{(d)}(\lambda)$, analytic in the half-plane $\Pi_N = \{\lambda \in \mathbb{C};\ \text{Re}\,\lambda > N\}$. The correspondence between F and $\tilde{F}^{(d)}$ is expressible in a two-fold way via formulae (1) and (3) (with Θ_t replaced by θ and the arguments ζ, ζ', t dropped out), namely:

i) the sequence $\left\{f_\ell^{(d)} = \tilde{F}^d(\ell)\ ;\ \ell \text{ integer}, \ell \geq N\right\}$ is related to $F_{|[-1,+1]}$ by Eqs.(1).

ii) $\tilde{F}_+^{(d)}$ and $\tilde{F}_-^{(d)}$ are related respectively to the jumps F_+ and F_- of iF across the cuts σ_+ and σ_- by Eqs.(3).

Inverse formulae expressing F in terms of $\tilde{F}$ (and implying properties i) and ii)) can be written as well.

A first rigorous version of this property, stated as a bijection between classes of analytic functions $F[\cos\theta]$ and $\tilde{F}(\lambda)$ (for the case $d = 3$, and for analytic functions F admitting a single cut σ_+ and a square-integrable behaviour at infinity and on this cut) was proved by the mathematicians Stein and Wainger[S,W] some time after the pioneering works of Froissart and Gribov. In the general method described here (see [B,V-1] for complete proofs), the central role played by Fourier-Laplace-type transformations will be emphasized.

For $d = 2$, the theorem relies on Fourier-Laplace analysis with respect to the complex variable θ, namely one defines $\tilde{F}(\lambda) = \int_\gamma e^{i\lambda\theta} F(\theta) d\theta$, where $F(\theta) = F[\cos\theta]$ and γ is a contour enclosing the cuts $\underline{\sigma}_+ = \{\theta = iv;\ v \geq v_+ > 0\}$, $\underline{\sigma}_- = \{\theta = iv + \pi;\ v \geq v_- > 0\}$ (corresponding to σ_+ and σ_- in the $\cos\theta$-plane), as indicated on Fig. 2 of the Appendix. Properties i) and ii) then follow from a simple contour-distortion argument. In fact, if F_+, F_- denote the jumps of i $F(\theta)$ across $\underline{\sigma}_+$ and $\underline{\sigma}_-$, $\tilde{F}_+$ and $\tilde{F}_-$ are the corresponding Laplace transforms $\tilde{F}_\pm(\lambda) = \int_{v_\pm}^{\infty} e^{-\lambda v} F_\pm(v) dv$ (analytic in Π_N since $|F_\pm(v)| \leq C^t e^{Nv}$ in view of the power-boundedness assumption on $|F|$); on the other hand the Fourier series of f is also obtained by noticing that for $\lambda = \ell$ integer, the integral that defines $\tilde{F}$ reduces to $\tilde{F}(\ell) = \int_{-\alpha}^{2\pi-\alpha} e^{i\ell\theta} f(\theta) d\theta = f_\ell$.

For $d > 2$, one can give a proof in the same spirit, one first applies to the function $F(\theta) = F[\cos\theta]$ a certain transformation $F \longrightarrow \hat{F}_d(\theta) = C^t e^{i\frac{d-2}{2}\theta} \mathcal{A}_d^{(c)} F(\theta)$ which preserves the analyticity and increase properties of F in the θ-plane, and one defines $\tilde{F}^{(d)}(\lambda)$ in terms of $\hat{F}_d$ by the *same* Fourier-Laplace integral as for the case $d = 2$, namely $\tilde{F}^{(d)}(\lambda) = \int_\gamma e^{i\lambda\theta} \hat{F}_d(\theta) d\theta$. In the previous definition of $\hat{F}_d$, $\mathcal{A}_d^{(c)} F(\theta)$ denotes the analytic function associated with F by the following Abel-type integral:

$$\mathcal{A}_d^{(c)} F(\theta) = \int_0^\theta F(\theta') \left[2(\cos\theta - \cos\theta')\right]^{\frac{d-4}{2}} \sin\theta' d\theta' \quad .$$

Inverse formulae which imply the bijective character of the transformation $F \longrightarrow \tilde{F}$ can be derived from the combination of:

a) a reciprocal Fourier-Laplace formula which exhibits $\hat{F}(\theta)$ [or $F(\theta)$ for $d = 2$] *in its analyticity domain, with its relevant growth property,* via Sommerfeld-Watson-type integrals $\int_{-\infty}^{\infty} \tilde{F}(N + i\nu) e^{-i(N+i\nu)(\theta\pm\pi)} / [\sin \pi(N + i\nu)] d\nu$ and

b) (for $d > 2$) reciprocal Abel-type transformations expressed by the formulae

$$F(\theta) = \mathrm{Const}^t \left(\frac{1}{\sin\theta} \frac{d}{d\theta}\right)^{\frac{d-2}{2}} \mathcal{A}_d^{(c)} F(\theta)$$

for d even, and

$$F(\theta) = \text{Const}^t \left(\frac{1}{\sin\theta}\frac{\mathrm{d}}{\mathrm{d}\theta}\right)^{\frac{d-1}{2}} \int_0^\theta \mathcal{A}_d^{(c)} F(\tau) \frac{\sin\tau \, \mathrm{d}\tau}{\sqrt{\cos\tau - \cos\theta}}$$

for d odd.

We shall now describe a geometrical presentation of the theorem considered (in d-dimensional complex space) which integrates the previous proof and actually gives the full meaning of it. This presentation provides the connection, *via analytic continuation,* between Fourier analysis on the sphere $\mathcal{S}_{d-1} \approx SO(d)/SO(d-1)$ and an appropriate realization of Fourier-Laplace analysis on the unit one-sheeted hyperboloid $X_{d-1} \approx SO(1,d-1)/SO(1,d-2)$ (introduced in [F.,V.]).

Analytic continuation takes place on a *complex* unit hyperboloid $X_{d-1}^{(c)}$ in $\mathbb{C}^d$ (represented as the set $\left\{z = \left(z^{(0)}, ..., z^{(d-1)}\right) \in \mathbb{C}^d;\ z^2 \equiv z^{(0)^2} - z^{(1)^2} - \ldots - z^{(d-1)^2} = -1\right\}$) which contains $\mathcal{S}_{d-1}$ and X_{d-1} as submanifolds of real type, namely $\mathcal{S}_{d-1} = X_{d-1}^{(c)} \cap \left(i\mathbb{R} \times \mathbb{R}^{d-1}\right)$ and $X_{d-1} = X_{d-1}^{(c)} \cap \mathbb{R}^d$. One then considers classes of functions which enjoy analyticity, power boundedness and invariance properties in a certain "cut-domain" D of $X_{d-1}^{(c)}$.

More specifically, the functions considered are supposed to be invariant under the little group (isomorphic to $SO(1,d-2)$) of a certain space-like vector of X_{d-1} which we choose to be the unit-vector along the $z^{(d-1)}$-axis of coordinates; since these functions only depend on $z^{(d-1)}$, we are led to put $z^{(d-1)} = \cos\theta$ and to denote them by $F[\cos\theta]$; this makes the connection with the previous one-variable presentation.

The analyticity domain D of these functions is the preimage of the cut-plane $\mathbb{C} \setminus \sigma_+ \setminus \sigma_-$ in $X_{d-1}^{(c)}$ (through the projection $z \longrightarrow z^{(d-1)}$). In particular, the sphere $\mathcal{S}_{d-1}$ is embedded in D, and projects onto the interval $[-1,+1]$ in the $z^{(d-1)}$-plane; the cuts σ_+ and σ_- are the images of subsets Σ_+ and Σ_- of X_{d-1}, defined respectively by the conditions $z^{(0)} > 0$, $z^{(d-1)} \geq \cosh v_+$ (i.e. $\theta = iv$, $v \geq v_+$) and $z^{(0)} < 0$, $z^{(d-1)} \leq -\cosh v_-$ (i.e. $\theta = iv + \pi$, $v \geq v_-$); these subsets have been pictured on Fig. 4 of the Appendix. If $\Sigma_+^{(c)}$ (resp. $\Sigma_-^{(c)}$) denotes the subset of $X_{d-1}^{(c)}$ defined by the condition $z^{(d-1)} \geq \cosh v_+$ (resp. $z^{(d-1)} \leq -\cosh v_-$), D is the "cut-domain" $X_{d-1}^{(c)} \setminus \left(\Sigma_+^{(c)} \cup \Sigma_-^{(c)}\right)$.

The jumps F_+, F_- of iF across σ_+, σ_- can now be considered as functions on X_{d-1} (depending only on the coordinate $z^{(d-1)} = \cosh v$ or $-\cosh v$) with supports contained respectively in Σ_+ and Σ_-.

For $d = 2$, $X_{d-1}^{(c)}$ reduces to a complex hyperbola parametrized by θ (i.e. $z^{(0)} = -i \sin\theta$, $z^{(1)} = \cos\theta$), and Fourier-Laplace analysis applies to the functions $F[\cos\theta] \equiv F(\theta)$ in the way described above.

For $d > 2$, a Laplace-type transformation on X_{d-1} (introduced in [F.V.] under the name of "spherical Laplace transformation" on X_2) can be applied to all functions F_+ with support in Σ_+ satisfying the appropriate invariance and power-boundedness conditions (and playing the role of the jump of F across Σ_+, or across Σ_- up to the symmetry $z \longrightarrow -z$, in the present framework).

This transformation, expressed in the one-variable approach by $F_+ \longrightarrow \tilde{F}_+^{(d)} = \int_{\cosh v_+}^{+\infty} Q_\lambda^{(d)} F_+$ (as in Eq.(3)), admits (from the [F.V.]-presentation) a more interesting alternative form which makes use of "horocyclic coordinates" on X_{d-1} (the latter correspond to parabolic sections of X_{d-1} parallel to a given light-like hyperplane). These coordinates, introduced from group-thereotical considerations, provide a geometrical interpretation for the occurrence of Abel-type kernels in the approach previously described. The residual Fourier-Laplace transformation (from $\hat{F}$ to $\tilde{F}$) in that approach now appears to be performed with respect to a variable θ which is the logarithm of a certain light-like coordinate of the point z varying in $X_{d-1}^{(c)}$. Precise equations and pictures that support this descriptive account can be found in the Appendix.

The advantage of this viewpoint on the theorem considered is that it relies, for any dimension d, on usual Fourier-Laplace analysis together with the exploitation of geometrical facts. The initial statement, expressed in terms of special functions $P_\lambda^{(d)}$ and $Q_\lambda^{(d)}$ can then be seen as a by-product of the geometrical study, including the definition of these functions via appropriate integral representations (see Eqs.(A.7), (A.13)).

A further study that can be done in this geometrical framework concerns convolution products that can be defined on X_{d-1} and on S_{d-1} and their mutual connection by analytic continuation in $X_{d-1}^{(c)}$.

For $d = 2$, the link between the convolution of 2π-periodic functions F on $\mathbb{R}$ and the convolution of functions $F_\pm$ with support in $\mathbb{R}^+$ can be established for the class of analytic functions $F(\theta)$ in $\mathcal{I}_+ \setminus \left\{ \underline{\sigma}_+ \cup \underline{\sigma}_-, \text{ mod. } 2\pi \right\}$, $\mathcal{I}_+$ denoting the upper half-plane (see [B,V-2] and a hint of the proof in the Appendix).

For $d = 3$, a generalization of the convolution on $\mathbb{R}^+$ has been given in [F.V.] under the name of "Volterra convolution on the one-sheeted hyperboloid". It extends immediately to the d-dimensional case, and applies to the class of functions

F_+ on X_{d-1}, *with support in* Σ_+, previously considered. If F_+ and G_+ are two such functions, their convolution-product on X_{d-1} denoted by $(F_+ \diamond G_+)(z)$ is defined by an integral on a compact region of X_{d-1}, determined by the supports of F_+ and G_+ and contained in a double-cone of the ambient space $\mathbb{R}^d$ of the form $\left\{z';\ z \gtrsim z' \gtrsim 0\right\}$ ($z \gtrsim 0$ denoting the light-cone order relation in $\mathbb{R}^d$, namely $z^{(0)} \geq \left[z^{(1)^2} + \cdots + z^{(d-1)2}\right]^{1/2}$).

For the functions F, G analytic in D, of the class considered above, the link between the various convolution products $F_\varepsilon \diamond G_{\varepsilon'}$, $\varepsilon, \varepsilon' = +$ or $-$, and the (more standard) convolution product $F * G$ on the sphere S_{d-1} can be established[BV.2], as for the case $d = 2$, by a procedure of contour distortion on $X_{d-1}^{(c)}$ (it even applies to more general classes of functions which do not satisfy the $SO(1, d-2)$-invariance condition). It turns out that $F*G$ is itself analytic in D and admits jump functions $[F*G]_+$ and $[F*G]_-$ (across Σ_+ and Σ_-) given in terms of $F_\pm$, $G_\pm$ by the following formulae (analogous to those obtained for the case $d = 2$; see Eq.(A.0)):

$$[F * G]_+ = F_+ \diamond G_+ + F_- \diamond G_- \ , \tag{15}$$

$$[F * G]_- = F_+ \diamond G_- + F_- \diamond G_+ \ , \tag{16}$$

or equivalently (by using the symmetrical and antisymmetrical combinations of $F_\pm$, $G_\pm$ as in section 1):

$$[F * G]^{(s)} = F^{(s)} \diamond G^{(s)} \quad \text{and} \quad [F * G]^{(a)} = F^{(a)} \diamond G^{(a)} \tag{17}$$

The main property of these convolution products is the fact that they are transformed into ordinary products in the Fourier variable λ. (For $d = 2$ it reduces to the usual theorems on Fourier transforms). The derivation of this property for the Volterra convolution on X_{d-1} (given by [F.V.] for $d = 3$) relies on group-theoretical considerations and involves the proof of a "product formula" for the second-kind Legendre functions $Q_\lambda^{(d)}$ (see Eq.(A.9)). The corresponding (more standard) property for the sequence of Fourier coefficients f_ℓ, g_ℓ of functions F, G on the sphere S_{d-1} (namely $F * G \longrightarrow f_\ell \cdot g_\ell$) can then be reobtained (for functions F, G analytic in D) by taking the restrictions of $\widetilde{F^{(s)} \diamond G^{(s)}}(\lambda) = \tilde{F}^{(s)}(\lambda) \cdot \tilde{G}^{(s)}(\lambda)$ and $\widetilde{F^{(a)} \diamond G^{(a)}}(\lambda) = \tilde{F}^{(a)}(\lambda) \cdot \tilde{G}^{(a)}(\lambda)$ to the respective sets of integers $\{\lambda = 2\ell\}$ and $\{\lambda = 2\ell + 1\}$ ($\ell \in \mathbb{N}$), and by applying property i) of the theorem and formulae (17).

2.2 A basic property of four-point functions in complex momentum space

We will now describe how and why the previous geometrical scenario applies to the four-point function $F(k_1, k_2; k_1', k_2')$ in the framework of axiomatic Q.F.T. (complete proofs will be given in [B,V-3]).

With each point (ζ, ζ', t) in $\Delta^{(e)}$ with $t < 0$ (see remark ii) at the end), we will associate a $(d-1)$-dimensional submanifold $\Omega_{(\zeta,\zeta',t)}$ of complex momentum space, parametrized by a vector z varying in a complex unit hyperboloid $X_{d-1}^{(c)}$. For all points $[k] \equiv ((k_i, k_i'); i = 1,2; k_1 + k_2 = k_1' + k_2') = [k](z)$ in $\Omega_{(\zeta,\zeta',t)}$, the corresponding Lorentz invariants ζ_i, ζ_i', t will take constant values, namely those of the given point in $\Delta^{(e)}$; the remaining Lorentz invariant $\cos\Theta_t$ will be equal to the coordinate $z^{(d-1)}$ of z.

More specifically, $\Omega_{(\zeta,\zeta',t)}$ is defined as follows. The "outgoing momenta" k_1', k_2' are kept fixed in such a way that the 2-plane π' spanned by them is the $\left(z^{(d-1)}, z^{(d)}\right)$-plane of coordinates, while the "total energy-momentum" $K = k_1' + k_2'$ (with $K^2 = t < 0$) is taken along the $z^{(d)}$-axis. On the other hand, the 2-plane π spanned by the "incoming momenta" k_1, k_2 is variable, although (since $K = k_1 + k_2$) it always contains the $z^{(d)}$-axis, and the "complex scattering angle" Θ_t is the angle between π and π'. The vector z which parametrizes $\Omega_{(\zeta,\zeta',t)}$ is then chosen to be a unit vector of π orthogonal to K, the corresponding unit vector $z' = z_0'$ of π' (orthogonal to K) being fixed along the $z^{(d-1)}$-axis; it follows from this choice that $\cos\Theta_t = z \cdot z_0' = z^{(d-1)}$; z varies in the complex hyperboloid $X_{d-1}^{(c)}$ $(z^2 = -1)$ of the subspace $\mathbf{C}^d$ orthogonal to the $z^{(d)}$-axis, and the parametrization of $\Omega_{(\zeta,\zeta',t)}$ is given by the following orthogonal decompositions of k_i (resp. k_i') in π (resp. π') :

$$k(z): \left[\begin{array}{ll} k_1 & = \rho z + \left(w + \frac{1}{2}\right) K \\ k_2 & = -\rho z - \left(w - \frac{1}{2}\right) K \end{array}\right. \qquad k': \left[\begin{array}{ll} k_1' & = \rho' z_0' + \left(w' + \frac{1}{2}\right) K \\ k_2' & = -\rho' z_0' - \left(w' - \frac{1}{2}\right) K \end{array}\right. \tag{18}$$

with $\rho^2 = \Lambda(\zeta_1, \zeta_2, t)/4t$, $\rho'^2 = \Lambda(\zeta_1', \zeta_2', t)/4t$ (Λ being defined as in Property I), and $w = \zeta_1 - \zeta_2/2t$, $w' = \zeta_1' - \zeta_2'/2t$.

The main remarks to be done are then:

i) the section of $\Omega_{(\zeta,\zeta',t)}$ by the euclidean subspace $(\mathrm{Re}[k]^{(0)} = 0, \mathrm{Im}\left[\vec{k}\right] = 0)$ is a sphere $S_{(\zeta,\zeta',t)}$ represented by the set $\left\{z \in S_{d-1} = X_{d-1}^{(c)} \cap \left(i\mathbb{R} \times \mathbb{R}^{d-1}\right)\right\}$.

ii) the section of $\Omega_{(\zeta,\zeta',t)}$ by the Minkowskian subspace is a one-sheeted hyperboloid represented by the set $\left\{z \in X_{d-1} = X_{d-1}^{(c)} \cap \mathbb{R}^d\right\}$.

When the point (ζ,ζ',t) varies in $\Delta^{(e)}$, the corresponding spheres $S_{(\zeta,\zeta',t)}$ generate (up to a global rotation) the whole euclidean subspace, which is known to be embedded in the axiomatic analyticity domain of $F^{[A][R][S]}$.

The image of this subspace in the Lorentz invariant variables, namely $\{(\zeta,\zeta',t) \in \Delta^{(e)}\} \times \{\cos\Theta_t \in [-1,+1]\}$ is then sufficient to define the partial waves f_ℓ of F (according to Eq.(1)). However, Property I relies on a much stronger result, namely the fact that F is analytic, for each (ζ,ζ',t) in $\Delta^{(e)}$, in the whole corresponding submanifold $\Omega_{(\zeta,\zeta',t)}$ except at the points of the latter which belong to the cuts $\Sigma^{(s)} = \left\{[k];\ s=(k_1-k_1')^2 \geq s_0\right\}$ and $\Sigma^{(u)} = \left\{[k];\ u=(k_1-k_2')^2 \geq u_0\right\}$. This domain $\Omega^{(\mathrm{cut})}_{(\zeta,\zeta',t)} = \Omega_{(\zeta,\zeta',t)} \backslash \Sigma^{(s)} \backslash \Sigma^{(u)}$ is represented in $X^{(c)}_{d-1}$ (via the parametrization (18)) by a cut-domain D of the form $D = X^{(c)}_{d-1} \backslash \Sigma^{(c)}_{+} \backslash \Sigma^{(c)}_{-}$ (as described in §2.1), whose projection onto the $\cos\Theta_t$-plane is the cut-plane $\mathbb{C}\backslash\sigma_+\backslash\sigma_-$.

The proof of the fact that F is analytic in each subset $\Omega^{(\mathrm{cut})}_{(\zeta,\zeta',t)}$ relies on the following result obtained in [B.E.G.] (and corresponding to a refinement of results of [A] [R] and [S]): F is analytic at all *real* points $[k]$ whose corresponding Lorentz invariants (ζ_i,ζ_i',s,t,u) lie below their spectral regions, and at all *complex* points $[k]$ such that:

a) k_1', k_2' are real, with corresponding invariants ζ_1',ζ_2',t below their spectral regions.

b) Im $k_1 = -$Im k_2 is a time-like vector.

We shall call O.V.E.E.D. (One-Vector Extrapolation of the Euclidean Domain) this part of the axiomatic analyticity domain of four-point functions.

One can in fact check that in each submanifold $\Omega_{(\zeta,\zeta',t)}$, O.V.E.E.D. yields a large enough region for justifying the analytic continuation of F in the whole domain $\Omega^{(\mathrm{cut})}_{(\zeta,\zeta',t)}$ (e.g. by using the Lorentz invariance of F). Now, by taking into account the existence of temperate bounds of the form $|F([k])| \leq C|[k]|^N$ in the Wightman-LSZ axiomatic framework, one concludes that in each submanifold $\Omega_{(\zeta,\zeta',t)}$, F satisfies the geometrical scenario of §2.1, which leads to a more comprehensive presentation of Property I.

Property II relies on a slight generalization of the previous analyticity property of F which states that F is also analytic in cut-domains $\hat{\Omega}^{(\mathrm{cut})}_{(\zeta,\zeta',t)}$ (with s- and u-cuts again) of submanifolds $\hat{\Omega}_{(\zeta,\zeta',t)}$ parametrized by $k = k(z)$, $k' = k(z')$, with z and z' varying in $X^{(c)}_{d-1}$ ($k(z)$ being the parametrization specified by the first two Eqs.(18)). $\hat{\Omega}^{(\mathrm{cut})}_{(\zeta,\zeta',t)}$ is in fact generated from $\Omega^{(\mathrm{cut})}_{(\zeta,\zeta',t)}$ by the action of complex Lorentz transformations (leaving the $z^{(d)}$-axis invariant) under which the analyticity domain of F is stable[E].

This property is necessary for exploiting the Bethe-Salpeter equation (6), which can be written in euclidean space as follows:

$$\begin{aligned} F(k(z);k(z_0')) &= G(k(z);k(z_0')) + \int_{\zeta''\in\Delta_t} \omega_d(\zeta'',t) \\ &\times \int_{z''\in S_{d-1}} F(k(z);k(z''))\, G(k(z'');k(z_0'))\, \alpha_d(z'') \end{aligned} \tag{19}$$

(ω_d and α_d denoting appropriate differential forms). In the latter, the subintegral with respect to z'' appears (for each fixed values of ζ,ζ'',ζ',t) as a convolution product on the sphere S_{d-1} (i.e. $F(\zeta,\zeta'',t;\bullet) * G(\zeta'',\zeta',t;\bullet)$ in the notations of §2.1). The analyticity properties of F and G respectively in $\hat{\Omega}^{(\mathrm{cut})}_{(\zeta,\zeta'',t)}$ and $\Omega^{(\mathrm{cut})}_{(\zeta'',\zeta',t)}$ allow one to apply the results on convolution products and on their Fourier-Laplace transforms (based on the Volterra convolution on X_{d-1} in the sense of [F.V.]) described above in §2.1, and therefrom to justify Property II.

Remarks

i) These results on convolution products of functions analytic in the domain D of $X^{(c)}_{d-1}$ (see Eqs.(15),(16),(17)) thus yield, in view of the previous analysis, a new derivation of the Bethe-Salpeter equations for absorptive parts (i.e. Eqs.(11), (12)) from the initial equation (6) (or (19)).

ii) At $t = 0$, the same analysis as above can be applied, by taking the limit $K \longrightarrow 0$ (from any space-like direction) of the previous geometrical situation, so that one has:

$$\tilde{F}(\zeta,\zeta';0,\lambda) = \lim_{t\longrightarrow 0} \tilde{F}(\zeta,\zeta';t,\lambda),$$

for all ζ,ζ' (with $\zeta_1 = \zeta_2 < 0$, $\zeta_1' = \zeta_2' < 0$) and λ in Π_N.

On the other hand, for $K = 0$, one can also exploit the special $SO(1,d)^{(c)}$ (instead of $SO(1,d-1)^{(c)}$)-invariance of the analyticity domain of F with respect to the complex vector $k_1 = -k_2$, and introduce the corresponding transform $\tilde{F}_0(\zeta,\zeta',\lambda)$ by replacing $Q^{(d)}_\lambda$ by $Q^{(d+1)}_\lambda$ in formula (3) (in this connection, see the work by Bjorken[Bj] and our related comment in section 3).

iii) Since the theorem stated in §2.1 expresses the equivalence between the analyticity and growth properties of $F[\cos\theta]$ and those of $\tilde{F}(\lambda)$, one can say that our Property I is actually *equivalent to* the analyticity and temperate growth of the four-point function $F(k_1,k_2;k_1',k_2')$ in O.V.E.E.D.

3. HISTORICAL COMMENTS, CONCLUDING REMARKS AND OUTLOOK

Since the sixties, it has been felt by various people that the introduction of complex angular momentum analysis in particle physics should find a conceptual justification by starting from a group representation viewpoint and in particular from the mathematical results of Bargmann[Ba], Gelfand[Ge], Harish-Chandra[HC] on the unitary reprsentations of the group $SO(1,2)$ (see e.g. [Ru] for a review of various works in this direction and in particular the works by M. Toller quoted therein).

On the other hand, in the S-matrix approach based on the ideas of maximal analyticity, it was realized that the Froissart-Gribov representation of partial waves was a powerful tool that could even be exploited in a general multiparticle formalism (see in particular the work by A. White[W] on the foundations of Reggeon calculus, based on the hypotheses of "many-particle dispersion relations" in the sense of H.P. Stapp[St]; for a general rewiew, see also [C]).

However, for the following reasons, it was difficult to perform the synthesis of these two aspects.

a) As it was shown by Helgason[H] (see also [V-1]), the mathematical results of [Ba], [Ge], [HC] actually contain (from a geometrical viewpoint) harmonic analysis on the *two-sheeted* hyperboloid; but in this framework, the corresponding Fourier-type transformation, which involves first kind Legendre functions $P_\lambda^{(d)}$ (instead of the second-kind functions $Q_\lambda^{(d)}$) *cannot* yield analytic functions of λ in half-planes Π_N as those encountered in the Froissart-Gribov representation.

b) The relevant geometrical aspect of the latter, namely the fact that at fixed t, the absorptive parts have their supports inside regions of the *one-sheeted* hyperboloid determined by the future-cone ordering relation (i.e. via spectral conditions) was not suggested in a natural way by the S-matrix analytic framework since the latter was more usually specified in terms of the Lorentz invariants (as e.g. in the Mandelstam representation) than in terms of complex energy-momentum vectors (as a matter of fact, the Mandelstam double spectral region corresponds to complex energy-momentum configurations which have no simple physical interpretation).

As a matter of fact, expansions on the one-sheeted hyperboloid were essentially introduced (although not in an explicit way) in a different set of works (see in particular [A,S]) in connection with multiperipheral Bethe-Salpeter-type equations; their starting point was the equations for crossed-channel absorptive parts (of the type of Eqs.(11) (12)) which were of current use in the perturbative field-theoretical approach (see e.g. [Go]). In particular, in the works of Abarbanel and Saunders[A,S], one can find a (semi-heuristic) derivation of the partial diagonalization of the previous integral equation based on the product formula for second-kind Legendre functions. Finally, it was only in [F,V] (after the investigations of [V-2]) that these properties were completely established through a systematic study of Laplace transforms on the one-sheeted hyperboloid and of "Volterra-convolutions".

Strangely enough, the connection (Froissart-Gribov type) between the Laplace transform introduced in these works and the off-shell partial waves had remained unnoticed (probably by lack of consideration for the underlying analyticity properties of the off-shell amplitude and for the link between the Bethe-Salpeter equation for the amplitude and those for the absorptive parts). A remarkable exception, however, is an earlier work by Bjorken[Bj] who, in the special situation $t = 0$ (requiring the use of the Gegenbauer polynomials and associated functions $Q_\lambda^{(4)}$, simpler to handle than the Legendre functions $Q_\lambda^{(3)}$, as it is the case for all *even* dimensions d) was able to derive the meromorphic structure of the transform $\tilde{F}_0(\zeta, \zeta', \lambda)$ (see §2.2, remark ii)) by starting from a euclidean approximate Bethe-Salpeter equation; in the latter, the kernel G was an arbitrary *finite* sum of perturbative amplitudes and the relevant analyticity and growth properties of this kernel were exploited through a general integral representation of Feynman amplitudes established by Nambu.

In the present work, we have shown that, at least in the off-shell region $\Delta^{(e)}$ of (ζ, ζ', t)-space, and for the case of scalar theories, a consistent set of properties, which integrate the various aspects described in the previous historical survey, are consequences of the general principles of Local Quantum Field Theory (in $(d+1)$-dimensional space-time, $d \geq 2$), namely

a) The existence of an *analytic interpolation* $\tilde{F}(\zeta, \zeta'; t, \lambda)$ *of the partial waves* $f_\ell(\zeta, \zeta', t)$ in a half-plane Π_N, related by Froissart-Gribov-type formulae to the (s- and u-channel) absorptive parts of the four-point function F.

b) The *(partially) diagonalized Bethe-Salpeter structure* of this interpolation $\tilde{F}$, implying possibly a meromorphic continuation of the latter with poles in the joint variables (λ, t) (in a larger half-plane $\Pi_{N'}$ of the λ-plane).

c) The interpretation of these results in terms of *Fourier-Laplace analysis* on a relevant type of orbit of the group $SO(1, d-1)$ (or $SO(1, d)$ for the case $t = 0$), namely the *one-sheeted hyperboloid*.

Moreover (see our remark iii) in §2.2), this set of properties is actually *equivalent* to a corresponding set of properties of F (including analyticity, temperate growth and Bethe-Salpeter structure) in a definite part, called O.V.E.E.D. in §2.2., of its axiomatic analyticity domain in complex momentum space.

By starting from these results, one can undertake at least two types of investigations, which concern respectively the group-theoretical aspect of these results and the largest region Δ in (ζ, ζ', t)-space for which the previous analytic and structural properties of F and $\tilde{F}$ remain valid.

The first type of investigation should reconsider the question of unitary representations of $SO(1, d-1)$, in connection with the one-sheeted hyperboloid, which should yield the relevant framework for understanding (by analytic continuation) the relationship between these representations and those of the compact group $SO(d)$. A deeper understanding of this question (namely at the *level of the groups* instead of *that of homogeneous spaces* considered throughout this work) would also allow one to give a general treatment of non-scalar cases of field theories. On the other hand, in connection with the group-theoretical aspect, the effect of the existence of $\tilde{F}_0\,(\zeta, \zeta'; \lambda)$ (corresponding to the special invariance property of the analyticity domain at $t = 0$) on the structural properties of $\tilde{F}\,(\zeta, \zeta'; t, \lambda)_{|t=0}$ deserves to be considered.

Concerning the second type of investigation, one can make the following comments on the unknown region Δ, at this preliminary stage.

i) Δ can be characterized as the largest region containing $\Delta^{(e)}$ such that (as a consequence of the QFT axioms) the four-point function $F[\zeta, \zeta'; t, \cos\,\Theta_t]$ is analytic and at most power-like behaved in a cut-plane of the variable $\cos\,\Theta_t$, with the additional condition that the corresponding cuts (i.e. the "s-cut" and "u-cut") should not contain the points $\cos\,\Theta_t = \pm 1$. This amounts to say that F should satisfy a dispersion relation in $\cos\,\Theta_t$; but we notice that the additional condition is violated if one sticks to the situation $t \leq 0$, since in particular for physical values of the masses ζ, ζ', the points $\cos\,\Theta_t = \pm 1$ lie (for $t \leq 0$) on the borders of the s- and u-channel physical regions, which are embedded in the corresponding s- and u-cuts. For the simplest equal-mass case, it is expected that the domain of dispersion relations (on the mass shell) obtained by A. Martin[Ma-2], deprived of the negative t axis (including in particular the cut-circle $|t| < 4\mu^2$, $\text{Arg}\ t \neq \pi$) should belong to Δ.

ii) Δ is contained in the axiomatic analyticity domain $\hat{\Delta}$ of the partial waves $f_\ell(\zeta,\zeta',t)$; in view of Eq.(1), $\hat{\Delta}$ is the set of all points (ζ,ζ',t) such that the four-point function $F[\zeta,\zeta';t,\cos\Theta_t]$ is defined and analytic on the interval $[-1,+1]$ of the $\cos\Theta_t$-plane, or on a suitably distorted path with end-points $(-1,+1)$. The fact that $\Delta \subset \hat{\Delta}$ clearly follows from this characterization. As it results from two-particle structure analysis[B], the intersection of $\hat{\Delta}$ with the mass-shell always contains a ramified neighbourhood of the two-particle region (e.g. $4\mu^2 \leq t < 9\,\mu^2$ for the equal mass case), from which a number of poles may have to be excluded; the latter will be either real poles in the physical sheet, corresponding to bound states involved in the spectrum of the theory, or unspecified complex poles in unphysical sheets corresponding to resonances (as shown in [B.I.], a two-sheeted or infinite-sheeted domain is produced according to whether $(d+1)$ is even or odd).

It is highly unprobable (except for some miracle of the non-linear program...) that such a domain above the t-channel threshold could be shown to belong to Δ, namely that one could justify in general, in the axiomatic Q.F.T. framework, the validity of the "actual" Froissart-Gribov representation of partial-waves in the two-particle physical region and in the unphysical sheet region where resonances may occur.

A more modest but interesting scope, however, would be to show that the domain Δ always contains, for a large set of values of (ζ,ζ') (including the mass-shell values, if possible) a *positive* interval in the energy-variable t, so that a range of possible bound-states of the theory might be included in Δ, which would thus allow Regge's concept of pole interpolation in the λ-plane to be operational in this general Q.F.T. framework.

It is clear that, for proving this property, one must carry out some off-shell counterpart of A. Martin's method[Ma−2] relying on positivity properties of the (off-shell) absorptive parts, since in the arguments of analytic continuation based purely on locality and spectrum (e.g. in the standard proofs of dispersion relations), one does not get rid of the restrictive condition $t \leq 0$ (for all values of ζ,ζ' considered).

To conclude, we can say that, despite the limitations to analytic continuation due to the generality of the QFT axiomatic framework, the structural properties established here in the initial domain $\Delta^{(e)}$ already justify our statement (in the introduction) that "Regge's philosophy is at least conceptually implied by the axiomatic framework of Q.F.T.". The extension of these structural properties to non-scalar cases and to a region of momentum space in the energy-range $t > 0$ would of course give a stronger support to that statement, even if this region lies below the scattering threshold, and corresponds to off-shell values of the mass variables. In fact, one can think that this theoretical framework might still exist for the case of theories of "fundamental fields" without asymptotic particles, but producing a purely discrete spectrum via a "confining-type" Bethe-Salpeter kernel; if the analytic structure of such a field theory is reasonably close to the conventional one (concerning at first, necessarily, the O.V.E.E.D. region), the set of bound states (corresponding to the observed particles) could then be associated, via the theoretical procedure described here, with interpolating polar singularities in complex (t, λ)-space.

APPENDIX

FOURIER-LAPLACE TRANSFORMATION ON CUT-DOMAINS OF THE COMPLEX HYPERBOLOID $X_{d-1}^{(c)}$

I. THE CASE $d = 2$

On the complex hyperbola $X_1^{(c)}$ ($z^{(0)} = -i \sin\theta$, $z^{(1)} = \cos\theta$, $\theta = u + iv$), one considers the domain $D = X_1^{(c)} \backslash \left(\Sigma_+^{(c)} \cup \Sigma_-^{(c)}\right)$ (Fig.1), whose representation in the θ-plane is the periodic cut-plane of Fig.2:

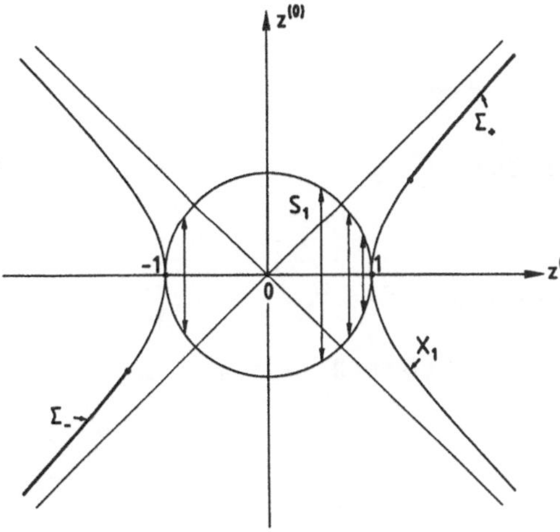

Figure 1

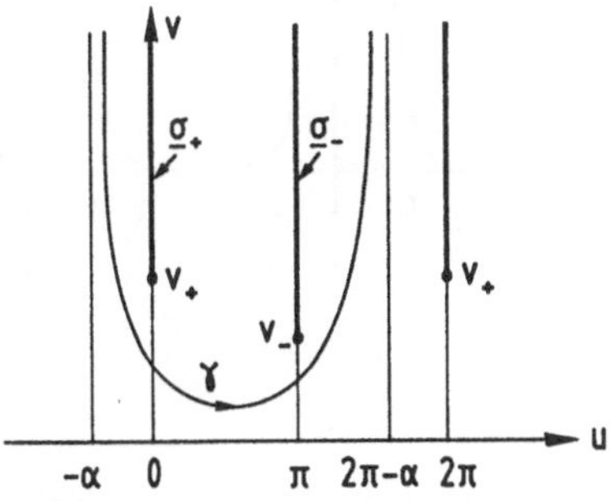

Figure 2

One then defines:

$$\tilde{F}(\lambda) = \int_\gamma e^{i\lambda\theta} F(\theta) d\theta,$$

with the prescription of Fig.2 for the contour γ.

Convolution product $F *^{(c)} G$ of functions F and G analytic in D:

Definition:

$$\left(F *^{(c)} G\right)(\theta) = \int_{\gamma_\theta} F(\theta') G(\theta - \theta') \, d\theta',$$

with the contour γ_θ (moving with θ) indicated on Fig. 3. When θ is real, γ_θ reduces to $[-\alpha, 2\pi - \alpha]$, which shows that $F *^{(c)} G$ is the analytic continuation of $F * G$ (in the cut-domain D).

Moreover the jump of $i\left(F *^{(c)} G\right)$ across the cut σ_+ (for example) can be computed by drawing γ_θ in the limiting positions corresponding to $\theta = \varepsilon + iv$ and to $\theta = -\varepsilon + iv$ (with $\varepsilon \longrightarrow 0$) and by taking the differences of the contributions of the "folded contours" thus obtained.

One gets easily convinced that this jump is given by

$$\left(F *^{(c)} G\right)_+ = F_+ \diamond G_+ + F_- \diamond G_-,$$

where

$$(F_\pm \diamond G_\pm)(v) = \int_0^v F_\pm(v') G_\pm(v - v') \, dv' . \qquad (A.0)$$

(Note that the threshold of the cut of $\left(F *^{(c)} G\right)_+$ is then $2 \min(v_+, v_-)$, instead of v_+)

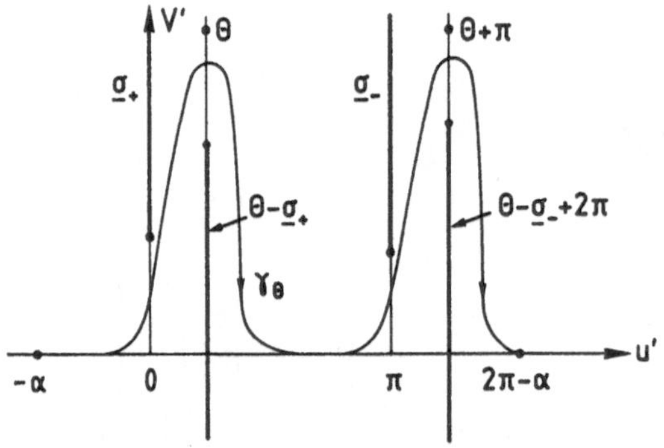

Figure 3

II. THE GENERAL CASE $d > 2$

a) Laplace-transformation on X_{d-1}

Two representations of $X_{d-1}^{(+)} = \left\{z \in X_{d-1};\ z^{(d-1)} \geq 1\right\}$:

- in polar coordinates: $z^{(0)} = \sinh w \cosh \varphi$, $z^{(d-1)} = \cosh w$

$$[\vec{z}] = \left(z^{(1)}, ..., z^{(d-2)}\right) = \sinh w \sinh \varphi [\vec{\alpha}], \quad [\vec{\alpha}] \in S_{d-3} \qquad (A.1)$$

- in horocyclic coordinates: $z^{(0)} = \sinh v + \frac{1}{2}\|\vec{x}\|^2 e^v$, $z^{(d-1)} = \cosh v - \frac{1}{2}\|\vec{x}\|^2 e^v$,

$$[\vec{z}] = \vec{x}\, e^v, \qquad \vec{x} \in \mathbb{R}^{d-2} \tag{A.2}$$

The sections $v = \text{const}^t$ are paraboloids in the planes $z^{(0)} + z^{(d-1)} = e^v$, called *horocycles.*

- For classes of functions $F_+(z) \equiv F_+[\cosh w]$ with support in $\Sigma_+ = \{z \in X_{d-1};\ z^{(d-1)} > \cosh v_+,\ z^{(0)} > 0\}$ ($\Sigma_+ \subset X_{d-1}^{(+)}$) the Laplace transform $\tilde{F}(\lambda)$ of F is defined as follows:

$$\tilde{F}_+(\lambda) = \int_{\Sigma_+} e^{-\lambda v} F_+[\cosh w] d\vec{x}\, dv \tag{A.3}$$

In the latter, it is convenient to use "mixed coordinates" v, w, $[\vec{x}]$ (see Fig.4); from (A.1), (A.2), one gets:

$$\vec{x} = e^{-v/2}[2(\cosh v - \cosh w)]^{1/2}\, [\vec{\alpha}], \quad \text{with} \quad [\vec{\alpha}] \in \mathcal{S}_{d-3},$$

which allows one to rewrite Eq.(A.3) as follows:

$$\tilde{F}_+(\lambda) = \omega_{d-2} \int_{v_+}^{\infty} e^{-(\lambda + \frac{d-2}{2})v} dv \int_{v_+}^{v} F_+[\cosh w][2(\cosh v - \cosh w)]^{\frac{d-4}{2}} \sinh w\, dw \tag{A.4}$$

or

$$\tilde{F}_+(\lambda) = \omega_{d-2} \int_{v_+}^{\infty} e^{-\lambda v} e^{-(\frac{d-2}{2})v} \mathcal{A}_d F_+(v) dv\ , \tag{A.5}$$

with

$$\mathcal{A}_d F_+(v) = \int_{v_+}^{v} F_+[\cosh w][2(\cosh v - \cosh w)]^{\frac{d-4}{2}} \sinh w\, dw \tag{A.6}$$

(ω_{d-2} denotes the area of $\mathcal{S}_{d-3}$).

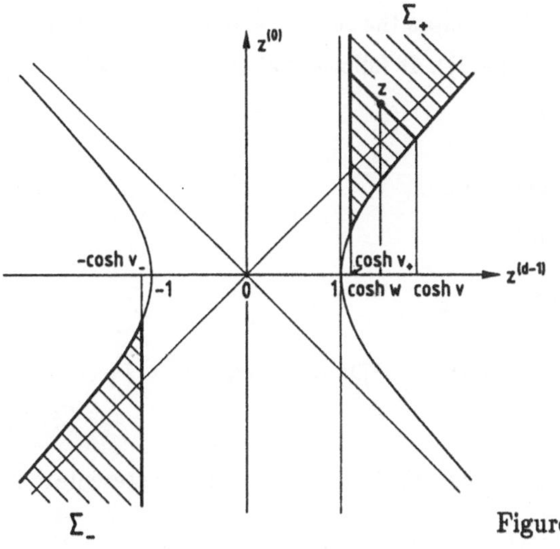

Figure 4

On the other hand, by introducing the second-kind function $Q_\lambda^{(d)}$ via the integral representation

$$Q_\lambda^{(d)}[\cosh w] = \frac{\omega_{d-2}}{(\sinh w)^{d-3}} \int_w^\infty \mathrm{e}^{-(\lambda+\frac{d-2}{2})v}[2(\cosh v - \cosh w)]^{\frac{d-4}{2}} \mathrm{d}v \qquad (A.7)$$

we obtain (by inverting the integrations in (A.4))

$$\tilde{F}_+(\lambda) = \int_{v_+}^\infty F_+[\cosh w] Q_\lambda^{(d)}[\cosh w](\sinh w)^{d-2} \mathrm{d}w \qquad (A.8)$$

The functions $Q_\lambda^{(d)}$ satisfy the following product formula

$$\begin{aligned} \omega_{d-2} \int_0^\infty Q_\lambda^{(d)} [\cosh v \cosh v' + \sinh v \sinh v' \cosh \xi] (\sinh \xi)^{d-3} \mathrm{d}\xi \\ = Q_\lambda^{(d)}[\cosh v] Q_\lambda^{(d)} [\cosh v'] \end{aligned} \qquad (A.9)$$

(used for the diagonalization of Volterra convolution on X_{d-1}).

b) The transformation $F \longrightarrow \tilde{F}$ for $F(z)$ analytic in the domain D of $X_{d-1}^{(c)}$

One defines $\tilde{F}(\lambda)$ by formulae similar to (A.5):

$$\tilde{F}(\lambda) = \omega_{d-2} \int_\gamma \mathrm{e}^{i(\lambda+\frac{d-2}{2})\theta} \mathcal{A}_d^{(c)} F(\theta) \mathrm{d}\theta \qquad (A.10)$$

with

$$\mathcal{A}_d^{(c)} F(\theta) = \int_{\gamma(0,\theta)} F[\cos \tau][2(\cos \theta - \cos \tau)]^{\frac{d-4}{2}} \sin \tau \, \mathrm{d}\tau, \qquad (A.11)$$

in the latter, the path $\gamma(0,\theta)$ with end points 0, and θ belongs to the domain $\mathbb{C}\backslash(\sigma_+ \cup \sigma_-)$; the contour γ in (A.10) is the same as for the case $d = 2$ (see Fig.2).

One checks that the jumps of $i\mathcal{A}_d^{(c)} F$ across the cuts σ_+ and σ_- are respectively $\mathcal{A}_d F_+$ and $\mathcal{A}_d F_-$ (defined by formula (A.6)). It follows that (as in the case $d = 2$) one has $\tilde{F}(\lambda) = \tilde{F}_+(\lambda) + \mathrm{e}^{i\pi\lambda}\tilde{F}_-(\lambda)$.

c) Passage to the "partial waves" on $\mathcal{S}_{d-1}$

For $\lambda = \ell$ integer, formula (A.10) yields

$$\begin{aligned} \tilde{F}(\ell) &= \omega_{d-2} \int_{-\pi}^{\pi} \mathrm{e}^{i(\ell+\frac{d-2}{2})u} \mathcal{A}_d^{(c)} F(u) \mathrm{d}u \\ &= \omega_{d-2} \int_0^\pi \sin t \, \mathrm{d}t \int_t^{2\pi-t} \mathrm{d}u \, \mathrm{e}^{i(\ell+\frac{d-2}{2})u}[2(\cos t - \cos u)]^{\frac{d-4}{2}} F[\cos t] \end{aligned} \qquad (A.12)$$

We can now introduce the polynomials $P_\ell^{(d)}$ via the integral representation (similar to Eq.(A.7) for $Q_\lambda^{(d)}$)

$$P_\ell^{(d)}(\cos t) = \frac{\omega_{d-2}}{(\sin t)^{d-3}} \int_t^{2\pi - t} e^{i(\ell + \frac{d-2}{2})u}[2(\cos t - \cos u)]^{\frac{d-4}{2}} du \qquad (A.13)$$

(which, by appropriate change of variables and contour distortion, is shown to be equivalent to the more familiar one:

$$P_\ell^{(d)}(\cos t) = \omega_{d-2} \int_0^\pi (\cos t + i \sin t \cos \alpha)^\ell (\sin \alpha)^{d-3} d\alpha)$$

Then in view of Eq.(A.13), we can rewrite Eq.(A.12) as follows

$$\tilde{F}(\ell) = \int_0^\pi F[\cos t] P_\ell^{(d)}(\cos t)(\sin t)^{d-2} dt,$$

which shows that $\tilde{F}(\ell) = f_\ell$.

REFERENCES

[A] H. Araki, J. Math. Phys. **2**, 163 (1961).

[A,S] H.D.I. Abarbanel and L.M. Saunders, Phys. Rev. **D2**, 711 (1970) and Annals of Physics **64**, 254 (1971).

[B] J. Bros, in "Analytic Methods in Math. Physics", R.P. Gilbert and R.G. Newton eds (New York, Gordon and Breach, 1970) p.85

[Ba] V. Bargmann, Ann. Math. **48**, 568 (1947).

[Bj] J.D. Bjorken, J. of Math. Phys. **5**, 192 (1964).

[B,E,G] J. Bros, H. Epstein, and V. Glaser, Nuovo Cim. **31**, 1265 (1964).

[B,I] J. Bros and D. Iagolnitzer, Phys. Rev. D**27**, 811 (1983) and Comm. Math. Phys. **88**, 235 (1983).

[B,L] J. Bros and M. Lassalle in "Structural Analysis of Collision Amplitudes", R. Balian and D. Iagolnitzer eds (Amsterdam, North Holland, 1976) p.97.

[B,V-1] J. Bros and G.A. Viano, "On the Connection between Harmonic Analysis on the Sphere and Harmonic Analysis on the One-Sheeted Hyperboloid", in preparation.

[B,V-2] J. Bros and G.A. Viano, "On the Connection between the Algebra of Kernels on the Sphere and the Volterra Algebra on the One-Sheeted Hyperboloid", in preparation.

[B,V-3] J. Bros and G.A. Viano, "Analyticity Properties with Respect to Complex Angular Momentum Variables in Axiomatic Quantum Field Theory", in preparation.

[C] P.D.B. Collins, "An introduction to Regge Theory and High Energy Physics" Cambridge University Press, 1977.

[E] H. Epstein, in "Axiomatic Field Theory", Chrétien and Deser eds, Gordon and Breach, New York 1966.

[F] M. Froissart, Phys. Rev. **123**, 1053 (1961).

[F,V] J. Faraut and G.A. Viano, J. Math. Phys. **27**, 840 (1986).

[G] V.N. Gribov, JETP **41**, 667 (1961). (Tr. Sov. Phys. JETP **14**, 478 (1962)).

[Ge] I.M. Gelfand; see [Vk] for a complete set of references.

[Go] M.L. Goldberger in "Multiperipheral Dynamics", Lectures at the Erice Summer School 1969.

[H] S. Helgason in "Lie Groups and Symmetric Spaces", Battelle Rencontre (C.M. De Witt and J.W. Wheeler eds), New York, Benjamin, 1968, p.1.

[HC] Harish Chandra, Amer. Journal of Math. **79**, p.193 and p.653 (1957).

[M] S. Mandelstam, Phys. Rev. **112**, 1344 (1958).

[Ma-1] A. Martin, Physics Letters **1**, 72 (1962).

[Ma-2] A. Martin, Nuovo Cim. **42**, 930 and **44**, 1219 (1966).

[R] D. Ruelle, Nuovo Cim. **19**, 356 (1961) and Thesis, Zürich 1959.

[Re] T. Regge, Nuovo Cim. **14**, 951 (1959).

[Ru] W. Rühl, "The Lorentz group and Harmonic Analysis", W.A. Benjamin Inc., New York 1970.

[S] O. Steinmann, Helv. Phys. Acta **33**, 257 (1960).

[St] H.P. Stapp, in "Structural Analysis of Collision Amplitudes", Balian and Iagolnitzer eds (Amsterdam, North Holland, 1976) p.277.

[S,W] E.M. Stein and S. Wainger, Arkiv für Matematik, Band 5, **37**, 553 (1964).

[S,Wi] R.F. Streater and A.S. Wightman, "PCT, Spin and Statistics and all that" Benjamin, New York 1964.

[V-1] G.A. Viano, Comm. Math. Phys. **26**, 290 (1972).

[V-2] G.A. Viano, Ann. Inst. H. Poincaré, **32**, 109 (1980).

[Vk] N.J. Vilenkin, "Special Functions and the Theory of Group Representations" Vol.22 (Translations of Mathematical Monographs), American Math. Soc., Providence, R.I. 1968.

[W] A.R. White, in "Structural Analysis of Collision Amplitudes", Balian and Iagolnitzer eds (Amsterdam, North Holland, 1976) p.433.

Inverse Scattering Revisited: Explicit Solution of the Marchenko-Martin Method

*N.N. Khuri**

Physics Department, The Rockefeller University,
1230 York Avenue, New York, NY 10021, USA

Abstract

We review Martin's inverse scattering method and Cornille's subsequent work on it's relation to the standard Gelfand -Levitan-Marchenko method.For the Yukawian class of potentials one obtains explicit solutions for the Marchenko equation,the potential, and the Jost functions.

I. Introduction

Often, a question posed by physicists presents the mathematician and the mathematical physicist with a problem whose resolution is of prime significance to both mathematics and physics. In the post war era, I can think of no better example of this than the case of inverse scattering. The physicist wanted a method to enable him to derive a unique potential from the scattering data (i.e. phase shifts) he measured. After some early confusion and abortive attempts, Bargmann[1] produced his famous counterexamples - different potentials which give exactly the same phase shift.

Within three years the problem was fully solved by the work of Gelfand and Levitan, and Marchenko[2]. The main result is as follows: The potential $V(r)$ is uniquely determined when one is given the three quantities: (a) the phase shift $\delta(E)$ for $0 < E < \infty$, (b) the number N and the energies $E_1, ..., E_N$ of the bound states, (c) a set of real, positive numbers $a_1^2, ..., a_N^2$, one corresponding to each bound state; where a_j is determined by the asymptotic behavior of the normalized bound state wave function, $u(r, E_j) \longrightarrow a_j e^{-\kappa_j r}$ as $r \longrightarrow \infty$, with $E_j = -\kappa_j^2$. An excellent review of this subject by Faddeev[3] first appeared in 1959. This almost closed the subject, at least for the original question of constructing a potential from a given $\delta(k)$. One exception is Dyson's[4] expression for $V(r)$ in terms of the Fredholm determinant of the Marchenko kernel.

---

* Work supported in part by the U.S. Department of Energy under Grant No. DOE-AC02-87ER-40325 Task B

There have of course been many extensions and applications which have been well covered in text books[5].

One important result not usually commented on in the standard literature on inverse scattering is the method due to Martin[6]. This method had a completely different origin. In the nineteen sixties particle theorists became interested in the class of potentials that have a representation $V(r) = \int_m^\infty C(\lambda)e^{-\lambda r}d\lambda$, the so-called Yukawian class of potentials. Scattering for these potentials was extensively studied especially in relation to Regge poles and the Mandelstam representation. As in the case of the pure Yukawa potential these potentials were mathematically similar to what one would get from the non- relativistic limit of a quantum field theory.

For potentials of the Yukawian class it was shown[7] that the partial wave scattering amplitudes, $a(E) \equiv e^{i\delta(E)} sin\delta(E)/2i\sqrt{E}$ have the following analyticity properties in E: (i) $a(E)$ was analytic in E, with two branch cuts on the real axis, one for $0 \leq E < \infty$, and the other for $-\infty < E \leq -m^2/4$, where m^{-1} is the range of $V(r)$, (ii) $|a(E)| \longrightarrow 0$ as $|E| \longrightarrow \infty$ in the cut plane, iii) $a(E)$ has simple poles at the bound state energies $E = -E_b$. It was also shown[8] that given the value of the discontinuity across the left hand cut, $q(E) \equiv [a(E + i0) - a(E - i0)]/2i$, for $-\infty < E \leq m^2/4$, one could determine $a(E)$ and the bound states via a Fredholm type equation.

Martin[6] developed an iterative scheme which enables one to reconstruct the Laplace transform of the potential from the discontinuity $q(E)$. He showed that if we are given $q(E)$ in the interval $\frac{m^2}{4} \leq -E \leq \frac{n^2m^2}{4}$, then in n steps of iteration we can find exactly $C(\lambda)$ for $m < \lambda < nm$. He gave the result for the first two iterations, and it looked that the general term would be rather complicated.

Following Martin's work, Gross and Kayser[9] considered the Marchenko equation for the Yukawian class of potentials. They noted that, for this class, the Marchenko kernel is a Laplace tansform of the discontinuity of $S(k)$ across the cut on the positive imaginary k-axis. They were the first to obtain Dyson's more general formula giving the potential in terms of the Fredholm determinant of the Marchenko kernel. Simultaneously, and independently Cornille [10] carried out an extensive analysis of the relation between Martin's and Marchenko's methods.He obtained the determinant formula and other interesting results.

In this book, honoring Martin's contributions it is timely to review this work, especially since many working on standard inverse scattering seem to be unaware of it. Note for example ref.4, page 162, second paragraph. We carry out this review using different methods. In addition we were able to obtain some new results such as equ.(4.43) where the Fredholm determinants are explicitly evaluated. We also give the exact expression for the Laplace weight, $C(\lambda)$, in terms of the left hand discontinuity of the S-matrix.

But our motivation for looking again at this problem goes beyond the main purpose of this book. It is connected to recent work by this author[11] on a physical scattering problem where the redundant poles of $S(k)$ are constructed to be the zeros of the Riemann zeta function, $\zeta(1/2+k)$. For more details the reader is referred to reference 11.

First we note that for the Yukawian class of $S(k)$, the Marchenko kernel is represented by a Laplace transform of $\nu(\alpha)$, the discontinuity of $S(k)$, on the pure imaginary axis, $k = i\alpha, m/2 < \alpha < \infty$. This fact simplifies the Marchenko equation and makes it explicitly solvable. We obtain an explicit solution of the Martin method giving $C(\lambda)$ in terms of $\nu(\alpha)$ for any interval $m \leq \lambda < M$ as a finite sum of simple functionals of $\nu(\alpha)$. One can also obtain an explicit expression for the two Jost solutions $f(k,x)$ and $f(-k,x)$ in terms of $\nu(\alpha)$, and verify directly that our expressions satisfy the Schrödinger equation.

Finally, we give necessary and sufficient conditions for an arbitrary real summable function, $\nu(\alpha)$, $m/2 < \alpha < \infty$, which guarantee the existence of a potential, $V(x)$, for which $\nu(\alpha)$ is the left hand cut related to $S(k)$.

In section II we review the Marchenko method, and in section III we give a brief review of Martin's method.The main results are in Section IV. In section V we consider the case of constructing potentials from arbitrary real functions, $\nu(\alpha)$.

II. The Marchenko Equation

For a large class of potentials, essentially those satisfying the condition $\int_0^\infty r|V(r)|dr < \infty$, the $l=0$ S-matrix, $S(k)$, has the following properties[3]:

$$(a) \quad |S(k)| = S(0) = S(\infty) = 1; \qquad S(k) = \overline{S}(-k) = \overline{S}(k)^{-1};$$

$$(b) \quad [S(k)-1] = \int_{-\infty}^{+\infty} F(x)e^{-ikx}dx,$$

and $F(x)$ is absolutely integrable.

$$(c) \quad argS(k=+\infty) - argS(k=-\infty) = -4i\pi N; \quad N \geq 0;$$

where N is the number of discrete eigenvalues.

The phase shifts are given by:

$$S(k) = e^{2i\delta(k)}. \tag{2.1}$$

Given an $S(k)$ satisfying conditions (a), (b), and (c); N positive numbers $\kappa_1, ..., \kappa_N$ determining the positions of the bound states, $E_j = -\kappa_j^2$; and N normalization constants, $a_1^2, ..., a_N^2$, one defines the Marchenko kernel as follows:

$$F(x) = \frac{1}{2\pi} \int_{-\infty}^{+\infty} [S(k) - 1] e^{ikx} dk - \sum_{n=1}^{N} a_n^2 e^{-\kappa_n x} \quad ,0 \leq x < \infty. \tag{2.2}$$

The function $F(x)$ is in $L_2(0, +\infty)$. The Marchenko equation is then given by

$$A(x,y) = F(x+y) + \int_x^\infty F(y+t) A(x,t) dt. \tag{2.3}$$

The existence of solutions for this integral equation has been established[3], and $A(x,y)$ is differentiable and summable for $0 \leq x \leq y < \infty$. In terms of $A(x,y)$, the potential is given uniquely by

$$V(x) = -2 \; \frac{d\, A(x,x)}{dx} \; , \quad 0 \leq x < \infty. \tag{2.4}$$

In addition $A(x,y)$ determines the Jost solutions of the Schrödinger equation as follows:

$$f^{(\pm)}(k,x) = e^{\mp ikx} + \int_x^\infty A(x,y) e^{\mp iky} dy, \tag{2.5}$$

where $f^{(\pm)}(k,x)$ are normalized such that $f^{(\pm)}(k,x) \longrightarrow e^{\mp ikx}$ as $x \longrightarrow +\infty$. The potential, $V(x)$, defined by (2.4) is guaranteed to have the S-matrix we started with, and N bound states at $k = i\kappa_j$, $j = 1, ..., N$. Furthermore, the normalized wave-functions for the bound states, u_j, satisfy $u_j(x, -\kappa_j^2) \longrightarrow a_j e^{-\kappa_j x}$ as $x \longrightarrow +\infty$ with the same a_j's that appear in (2.2).

A necessary and sufficient condition for the validity of

$$\int_0^\infty x |V(x)| dx \quad < \infty, \tag{2.6}$$

is that

$$\int_0^\infty x |F'(2x)| dx \quad < \infty. \tag{2.7}$$

Finally, we stress the fact that since F is a Fredholm kernel the existence of a solution of (2.3) necessarily implies that solutions of the homogeneous equation,

$$A(x,y) = \int_x^\infty F(y+z)A(x,z)\,dz, \tag{2.8}$$

do not exist for any real x, $0 \le x < \infty$.

III. Martin's Method

We give a brief review of Martin's method for S-waves.

One starts with the class of potentials which can be represented as:

$$V(x) = \int_m^\infty C(\lambda)e^{-\lambda x}d\lambda, \quad 0 \le x < \infty, \; m > 0. \tag{3.1}$$

The inverse Laplace transform of $V(x)$, $C(\lambda)$, is restricted to satisfy:

$$\int_m^\infty |C(\lambda)|\frac{d\lambda}{\lambda^2} < \infty. \tag{3.2}$$

This last condition guarantees that $x|V(x)|$ is integrable at $x = 0$. Hence we are dealing with potentials less singular than x^{-2} as $x \longrightarrow 0$. For $x > 0$, (3.1) guarantees that $V(x)$ is bounded.

Next one considers the Jost solutions of the Schrödinger equation, $f^{(\pm)}(k,x)$, where

$$f^{(\pm)}(k,x) \longrightarrow e^{\mp ikx} \quad as \quad x \longrightarrow +\infty; \tag{3.3}$$

and

$$\frac{d^2 f^{(\pm)}}{dx^2}(k,x) + k^2 f^{(\pm)}(k,x) = V(x)f^{(\pm)}(k,x). \tag{3.4}$$

It is convenient to define solutions $h_\pm(k,x)$ which are normalized to unity as $x \longrightarrow +\infty$, e.g.

$$f^{(+)}(k,x) = h_+(k,x)e^{-ikx}; \qquad f^{(-)}(k,x) = h_-(k,x)e^{ikx}. \tag{3.5}$$

$h_\pm(k,x) \longrightarrow 1$ as $x \longrightarrow +\infty$ for all real k, $-\infty < k < +\infty$. For $h_\pm$, Schrödinger's equation (3.4) becomes,

$$\frac{d^2 h_\pm}{dx^2} \mp 2ik\frac{dh_\pm}{dx} = V(x)h_\pm. \tag{3.6}$$

Martin writes for $h_\pm(k,x)$ the Ansatz:

$$h_\pm(k,x) = 1 + \int_m^\infty \rho_\pm(k,\alpha)e^{-\alpha x}d\alpha. \tag{3.7}$$

Substituting (3.1) and (3.7) in equation (3.6) we get:

$$\alpha(\alpha \pm 2ik)\rho_\pm(k,\alpha) = C(\alpha) + \theta(\alpha - 2m)\int_m^{\alpha-m} C(\alpha-\beta)\rho_\pm(k,\beta)d\beta. \tag{3.8}$$

The limits of integration are determined by the fact $C(\alpha) = 0$ and $\rho_\pm(k,\alpha) = 0$ for $\alpha < m$. Equation (3.8) can be solved by iteration with a finite number of steps. Setting $\rho^0_\pm(k,\alpha) \equiv C(\alpha)$ one can easily show that $\rho^{(n)}_\pm(k,\alpha) = 0$ if $\alpha < (n+1)m$, where $\rho^{(n)}_\pm$ is obtained from $\rho^{(n-1)}_\pm(k,\alpha)$ by substituting the latter on the right hand side of (3.8),

$$\rho^{(n)}_\pm(k,\alpha) = \frac{1}{\alpha(\alpha \pm 2ik)} \int_m^{\alpha-m} C(\alpha-\beta)\rho^{(n-1)}_\pm(k,\beta)d\beta. \tag{3.9}$$

For any finite α, it is clear that $\rho_\pm(k,\alpha)$ will have singularities in k only on the imaginary k-axis, with those of ρ_+ having $\mathrm{Im}k > 0$ and those of $\rho_-(k,\alpha)$ with $\mathrm{Im}k < 0$. The Jost functions are given by

$$h_\pm(k,0) \equiv f(\pm k) = 1 + \int_m^\infty \rho_\pm(k,\alpha)\, d\alpha. \tag{3.10}$$

Martin proved the uniform and absolute convergence of (3.10) for potentials satisfying conditions (3.1) and (3.2). Thus $h_+(k,0) = f(k)$ is analytic in the cut k-plane with a branch cut on the positive imaginary k-axis, $\frac{m}{2} \leq \mathrm{Im}k < \infty$; and $h_-(k,0) \equiv f(-k)$ has a cut on the negative k-axis. In addition one obtains $\lim_{k\to\infty} f(k) = 1$ along any direction in the k-plane.

Equation (3.8) is the equivalent of the Schrödinger equation, given that one can construct $h_\pm(k,x)$ from $\rho_\pm(k,x)$. In addition to (3.8), Martin derived a new equation relating $C(\alpha)$, $\rho_\pm(k,\alpha)$, and the discontinuity of $S(k)$ along the imaginary axis, $\frac{m}{2} \leq \mathrm{Im}k < \infty$, denoted by $\nu(\alpha)$, for $k = i\alpha$.

For potentials satisfying (3.1) and (3.2) we have just noted that $S(k) \equiv h_+(k)/h_-(k)$, is analytic for $\mathrm{Im}k > 0$ with a branch cut on the imaginary axis starting at $\mathrm{Im}k = m/2$. The bound states appear as poles of $S(k)$ resulting from the zeros of $h_-(k)$, which again fall on the imaginary axis. We define the discontinuity $\nu(\alpha)$, as

$$\frac{1}{2i}[S(k = i\alpha + \epsilon) - S(k = i\alpha - \epsilon)] \equiv \nu(\alpha), \quad \frac{m}{2} \leq \alpha < \infty. \tag{3.11}$$

We set $\nu(\alpha) \equiv 0$ for $\alpha < m/2$, and the new equation of Martin is:

$$-\frac{2\lambda\nu(\lambda)}{\pi} = C(2\lambda) + \int_m^{2\lambda-m} C(2\lambda-\alpha)\rho_+(i\lambda,\alpha)d\alpha. \tag{3.12}$$

From this equation and (3.8) one can iteratively reconstruct $C(\lambda)$ from $\nu(\alpha)$. Note first that for $m/2 \leq \lambda < m$, we have

$$C(2\lambda) = -2\lambda\frac{\nu(\lambda)}{\pi}\ , \quad \frac{m}{2} \leq \lambda < m. \tag{3.13}$$

Obviously for $\lambda < m$ the integral in (3.12) vanishes. With $C(\xi)$ known for $m \leq \xi < 2m$ from (3.13), we can substitute in (3.8) and obtain $\rho_-(i\lambda, \alpha)$ for $m \leq \alpha < 2m$ which we can now use in (3.12) to get $C(\xi)$ in a larger interval,

$$C(2\lambda) = -2\frac{\lambda\nu(\lambda)}{\lambda} + \frac{4}{\pi}\int_m^{2\lambda-m} \nu(2\lambda - m)\nu(\alpha)d\alpha \; ; \quad m/2 < \lambda < m.$$

This procedure can be continued and given $\nu(\alpha)$ in the interval $m/2 \leq \alpha \leq \frac{1}{2}nm$, we can obtain $C(\lambda)$ for $m \leq \lambda < nm$, where n is a positive integer. the potential is the Laplace transform of $C(\lambda)$ and definitely converges for $x > 0$.

IV. Solution of the Marchenko Equation for Yukawian Potentials

In this section we rederive some of the results of refs.9 and 10, in addition we obtain additional results.

We start by defining a class of $S(k)$ which we call the Yukawian class: *Definition*: The S-matrix, $S(k)$, is Yukawian, if in addition to the standard properties (a), (b), and (c) given at the beginning of section II, it has the following two additional properties: (d) $S(k)$ is analytic in k for Im$k > 0$, except on the positive imaginary axis where $S(k)$ has a branch cut extending on the interval $m/2 \leq \text{Im}k < \infty$, Re$k = 0$, and simple poles corresponding to the bound states at $k = i\kappa_n, \kappa_n > 0$, $n = 1, ..., N$. (e) For large $|k|$, Im$k \geq 0$, we have the asymptotic property[12], valid in all directions,

$$|S(k) - 1| \leq const.|k|^{-\delta} \;, \quad 0 < \delta < 1, \; \text{Im}k \geq 0, \tag{4.1}$$

for all k such that $|k| > M$, where M is a large constant.

The first thing to note is that when $S(k)$ belongs to the Yukawian class it is analytic and the normalization constants a_n^2 that appear in the definition (2.2) of the Marchenko kernel, $F(x)$, are just the residues of $S(k)$ for the bound state pole at $k = i\kappa_n, \kappa_n > 0$. Namely, we have

$$\frac{1}{2\pi}\oint_{C_n} S(k)dk \equiv a_n^2 \;; \tag{4.2}$$

where the contour C_n is a small circle that encloses the pole at $k = i\kappa_n$ in the positive sense. Previously we defined a_n through the asymptotic behavior of the normalized bound state wave function, $u(r, \kappa_n)$, which as $r \longrightarrow \infty$ behaves as $u(r, \kappa_n) \longrightarrow a_n e^{-\kappa_n r}$, and $\int_0^\infty u^2(r, \kappa_n)dr = 1$. Second, one should note that the discontinuity, $\nu(\alpha)$, of $S(k)$ along the imaginary k-axis, defined as in eq.(3.11), is summable and satisfies

$$\int_{\frac{m}{2}}^{\infty} \frac{|\nu(\alpha)|}{\alpha} d\alpha < \infty . \tag{4.3}$$

This is essentially guaranteed by the bound (4.1). In some cases $\nu(\alpha)$ is a distribution and (4.3) is then appropriately modified by smearing it in the standard way.

We now prove the following lemma:

Lemma1: *Given an $S(k)$ belonging to the Yukawian class, then the Marchenko kernel, $F(x)$, defined by equ. (2.2) is given by the Laplace transform of, $\nu(\alpha)$, the discontinuity of $S(k)$ across the cut along the imaginary axis, $k = i\alpha$, $m/2 \leq \alpha < \infty$,*

$$F(x) = -\frac{1}{\pi} \int_{\frac{m}{2}}^{\infty} \nu(\alpha) e^{-\alpha x} d\alpha , \quad 0 < x < \infty. \tag{4.4}$$

This result holds even in the presence of bound states.

Proof:

From (2.2) we have:

$$F(x) = \frac{1}{2\pi} \int_{-\infty}^{+\infty} [S(k) - 1] e^{ikx} \quad - \sum_{n=1}^{N} a_n^2 e^{-\kappa_n x} \quad ; \quad 0 \leq x < \infty.$$

We can deform the contour of integration from that along the real k-axis, to a contour surrounding the branch cut, i.e.

$$F(x) = \frac{1}{2\pi} \int_C [S(k) - 1] e^{ikx} dk , \tag{4.4a}$$

where C starts at $(-\epsilon, +i\infty)$ and descends to $(-\epsilon, +i\frac{m}{2})$, turns around the point $k = i\frac{m}{2}$, and then extends from $(+\epsilon, +i\frac{m}{2})$ to $(+\epsilon, +i\infty)$. It follows from (4.2) that the contribution from the residues of the poles of $S(k)$ exactly cancels the second term in (2.2). Thus the information about the bound states is no longer explicitly showing in (4.5a). Furthermore, the bound (4.1) on $|S(k) - 1|$ for large $|k|$ guarantees that as we shift the contour there will be no contribution from the large semicircle. Noting the definition,(3.11),

$$\nu(\alpha) = \frac{1}{2i} [S(i\alpha + \epsilon) - S(i\alpha - \epsilon)], \quad m/2 \leq \alpha < \infty,$$

equation (4.4a) becomes

$$F(x) = -\frac{1}{\pi}\int_{\frac{m}{2}}^{\infty} \nu(\alpha)e^{-\alpha x}d\alpha \quad . \tag{4.4}$$

This integral is obviously convergent for any x with $Re x > 0$, see (4.1). For $\nu(\alpha) \approx O(\alpha^{-\delta})$ as $\alpha \longrightarrow \infty$ we see that $F(x) = O(x^{\delta-1})$ near $x \approx 0$ with $0 < \delta < 1$. (Thus with $1 > \delta > 0$ we have only potentials, $V(x)$, which are less singular than x^{-2} at the origin, and $\int_0^c x|V(x)|dx$ exists for any $c > 0$). This completes the proof of Lemma 1.

The fact that the bound state information does not explicitly appear in (4.4) is not mysterious. One knows, for Yukawian potentials, that given $\nu(\alpha)$ one can obtain the Jost functions from the so-called N/D method via a Fredholm equation[8]. The bound states will appear as the zeros of $f(-k)$. Hence $\nu(\alpha)$ has in it all the information about the bound state poles and their residues. This fact will become even more apparent below.

With $F(x)$ expressed as in (4.4) we consider the Marchenko equation (2.3),

$$A(x,y) = F(x+y) + \int_x^{\infty} F(y+z)A(x,z)dz.$$

For $S(k)$ belonging to the Yukawian class we write an Ansatz for $A(x,y)$ as:

$$A(x,y) = \int_{\frac{m}{2}}^{\infty} d\sigma \int_{\frac{m}{2}}^{\infty} d\alpha \ \tilde{A}(\sigma,\alpha)e^{-\sigma x}e^{-\alpha y}. \tag{4.5}$$

Substituting (4.4) and (4.5) in (2.3) we get

$$\tilde{A}(\sigma,\alpha) = -\frac{\nu(\alpha)}{\pi}\delta(\sigma-\alpha) - \frac{\nu(\alpha)}{\pi}\int_{\frac{m}{2}}^{\sigma-\alpha} d\alpha' \frac{\tilde{A}(\sigma-\alpha-\alpha',\alpha')}{\alpha'+\alpha}. \tag{4.6}$$

This can be considered as the Laplace transform of the Marchenko equation. Equation (4.6) can be simplified further by defining $\tilde{B}(\sigma,\alpha)$ as:

$$\tilde{A}(\sigma,\alpha) \equiv -\frac{\nu(\alpha)}{\pi}\tilde{B}(\sigma,\alpha). \tag{4.7}$$

We obtain for $\tilde{B}$

$$\tilde{B}(\sigma,\alpha) = \delta(\sigma-\alpha) - \frac{1}{\pi}\int_{\frac{m}{2}}^{\sigma-\alpha} d\alpha' \frac{\nu(\alpha')\tilde{B}(\sigma-\alpha-\alpha',\alpha')}{\alpha'+\alpha}. \tag{4.8}$$

It is obvious from both (4.6) and (4.7) that

$$\tilde{A}(\sigma,\alpha) \equiv 0 \ , \quad \sigma < \alpha, \tag{4.9}$$

and

$$\tilde{B}(\sigma,\alpha) \equiv 0 \ , \quad \sigma < \alpha.$$

In addition $\tilde{B}(\sigma,\alpha) = \tilde{A}(\sigma,\alpha) = 0$ if $\sigma < m/2$ where $\nu(\alpha)$ is defined such that $\nu(\alpha) = 0$ for $\alpha < m/2$.

The remarkable fact now is that for any finite region $\frac{m}{2} \le \alpha < \sigma < M$, $\tilde{B}(\sigma,\alpha)$ is exactly determined by a finite number of iterations of (4.8). Iterating (4.8) N times we get

$$\tilde{B}(\sigma,\alpha) = \sum_{n=0}^{N} \tilde{B}_n(\sigma,\alpha) + \tilde{R}_{N+1}(\sigma,\alpha), \tag{4.10}$$

where

$$\tilde{B}_n(\sigma,\alpha) = (-\frac{1}{\pi})^n \int_{\frac{m}{2}}^{\infty} d\alpha_1 \cdots \int_{\frac{m}{2}}^{\infty} d\alpha_n \frac{\prod_{j=1}^{n} \nu(\alpha_j)}{\prod_{j=0}^{n-1}(\alpha_j + \alpha_{j+1})} \delta(\sigma - \alpha - 2\sum_{j=1}^{n} \alpha_j), \tag{4.11}$$

with $n \ge 1$, $\alpha_0 = \alpha$ and

$$\tilde{B}_0(\sigma,\alpha) = \delta(\sigma - \alpha). \tag{4.12}$$

Note that the upper limits of integration in (4.11) can be taken to infinity because of the presence of the δ-functions. Second we note that

$$\tilde{B}_n(\sigma,\alpha) \equiv 0 \quad for \quad (\sigma - \alpha) < nm, \tag{4.13}$$

given that $\alpha_j \ge \frac{m}{2}$.

The function $R_{N+1}(\sigma,\alpha)$ in (4.10) is given by

$$R_{N+1}(\sigma,\alpha) = (-\tfrac{1}{\pi})^{N+1} \int_{\frac{m}{2}}^{L_1} d\alpha_1 \cdots \int_{\frac{m}{2}}^{L_{N+1}} d\alpha_{N+1}$$

$$\frac{(\prod_{j=1}^{N+1} \nu(\alpha_j))\tilde{B}(\sigma - \alpha - 2\sum_{j=1}^{N} \alpha_j - \alpha_{N+1}, \alpha_{N+1})}{(\prod_{j=0}^{N}(\alpha_j + \alpha_{j+1}))}, \tag{4.14}$$

where again $\alpha_0 = \alpha$, and

$$L_j = (\sigma - \alpha - 2\sum_{l=1}^{j-1} \alpha_l). \tag{4.15}$$

We see now that $\tilde{B}$ appearing on the right in (4.14) will vanish unless

$$\sigma - \alpha - 2\sum_{j=1}^{N+1} \alpha_j \ge 0 .$$

Hence, $R_{N+1}(\sigma,\alpha)$ will vanish if

$$(\sigma - \alpha) < (N+1)m.$$

Thus for any finite large M, and σ and α such that,

$$\frac{m}{2} \le \alpha \le \sigma < M , \tag{4.16}$$

$$\tilde{B}(\sigma,\alpha) = \sum_{n=1}^{N(M)} \tilde{B}_n(\sigma,\alpha) + \delta(\sigma-\alpha) \ , \tag{4.17}$$

where $N(M)$ is an integer defined by

$$Nm \leq M \leq (N+1)m. \tag{4.18}$$

The Laplace transform, $\tilde{A}(\sigma,\alpha)$, is:

$$\tilde{A}(\sigma,\alpha) = \sum_{n=0}^{N(M)} \tilde{A}_n(\sigma,\alpha) \ , \quad \frac{m}{2} < \alpha \leq \sigma < M, \tag{4.19}$$

with $\tilde{A}_0(\sigma,\alpha) = -\frac{\nu(\alpha)}{\pi}\delta(\sigma-\alpha)$, and $\tilde{A}_n(\sigma,\alpha) = -\frac{\nu(\alpha)}{\pi}\tilde{B}_n(\sigma,\alpha), \quad n \geq 1$. From equation (2.4) we have,

$$V(x) = -2\frac{d}{dx}(A(x,x)) = +2\int d\sigma \int d\alpha(\sigma+\alpha)\tilde{A}(\sigma,\alpha)e^{-(\sigma+\alpha)x} \ ,$$

and using the representation, $V(x) = \int_m^\infty C(\lambda)e^{-\lambda x}d\lambda$, we obtain

$$C(\lambda) = 2\lambda \int_{\frac{m}{2}}^{\lambda} d\alpha \tilde{A}(\lambda-\alpha;\alpha). \tag{4.20}$$

Substituting (4.19) and (4.11) we get an explicit expression for $C(\lambda)$ in terms of $\nu(\alpha)$ for $\lambda \leq M$:

$$C(\lambda) =$$

$$2\lambda \sum_{n=0}^{N(M)} (\frac{-1}{\pi})^{n+1} \int_{\frac{m}{2}}^{\infty} d\alpha_0 \cdots \int_{\frac{m}{2}}^{\infty} d\alpha_n \frac{(\prod_{j=0}^{n} \nu(\alpha_j))}{(\prod_{j=0}^{n-1}(\alpha_j+\alpha_{j+1}))} \delta(\lambda - 2\sum_{j=0}^{n} \alpha_j). \tag{4.21}$$

This gives $C(\lambda)$ exactly in the interval $m \leq \lambda < M < (N(M)+1)m$.

Equation (4.21) provides us with an explicit solution of the Martin problem. The result has simplicity and symmetry which was not apparent in the original Martin procedure. Of course one can easily check up to $n = 2$ with Martin's results.

To obtain $V(x)$ from $C(\lambda)$ we have to integrate over the whole interval $\frac{m}{2} \leq \lambda \leq \infty$, and (4.21) becomes a series whose convergence has to be checked. This can be easily done for all $x > x_0 > 0$, where x_0 is determined below.

From (4.3) we write

$$A \equiv \frac{1}{\pi}\int_{\frac{m}{2}}^{\infty}\frac{|\nu(\alpha)|}{\alpha}d\alpha < \infty. \tag{4.22}$$

We define x_0 as

$$x_0 = \frac{\log A}{m}. \tag{4.23}$$

For $x > x_0$, we can take the Laplace transform of $C(\lambda)$ term by term and the resulting series for $V(x)$ will be absolutely and uniformly convergent. We get for $x > x_0 > 0$:

$$V(x) = 4\sum_{n=0}^{\infty}\left(\frac{-1}{\pi}\right)^{n+1}\int_{\frac{m}{2}}^{\infty}d\alpha_0\cdots\int_{\frac{m}{2}}^{\infty}d\alpha_n\frac{\left(\prod_{j=0}^{n}\nu(\alpha_j)e^{-2\alpha_j x}\right)}{\left(\prod_{j=0}^{n-1}(\alpha_j+\alpha_{j+1})\right)}\left(\sum_{j=0}^{n}\alpha_j\right). \tag{4.24}$$

The absolute convergence for $x > x_0$ of this series is easy to check, and indeed it also converges for complex x provided $\mathrm{Re}x > x_0 \equiv \frac{\log A}{m}$. In the cases where $A < 1$, then $\log A < 0$ and (4.24) gives the potential for the whole interval $0 \le x < \infty$. Below we shall show how to obtain an explicit expression for $V(x)$ for all x, $0 \le x < \infty$, even when $A > 1$.

Before we do that we write down the Jost solutions $h_{\pm}(k,x)$ in terms of $\nu(\alpha)$. From (2.5), noting the definition $h_{\pm} = e^{\pm ikx}f_{\pm}$, we get

$$h_{\pm}(k,x) = 1 + \int_{x}^{\infty}A(x,y)e^{\pm ik(x-y)}dy. \tag{4.25}$$

Substituting the Ansatz (4.5) for $A(x,y)$ we obtain,

$$h_{\pm}(k,x) = 1 + \int_{m}^{\infty}d\lambda e^{-\lambda x}\left[\int_{\frac{m}{2}}^{\lambda}d\alpha\frac{\tilde{A}(\lambda-\alpha,\alpha)}{[\alpha \pm ik]}\right]. \tag{4.26a}$$

The Laplace weights $\rho_{\pm}(k,\lambda)$ defined in (3.7) are thus given by:

$$\rho_{\pm}(k,\lambda) = \int_{\frac{m}{2}}^{\lambda}d\alpha\frac{\tilde{A}(\lambda-\alpha,\alpha)}{[\alpha \pm ik]} \tag{4.26b}$$

Next inserting the solution (4.19) for $\tilde{A}$ we get:

$$h_{\pm}(k,x) = 1 + \sum_{n=0}^{\infty}\left(\frac{-1}{\pi}\right)^{n+1}\int_{\frac{m}{2}}^{\infty}d\alpha_0\cdots\int_{\frac{m}{2}}^{\infty}d\alpha_n\frac{\left(\prod_{j=0}^{n}\nu(\alpha_j)e^{-2\alpha_j x}\right)}{\left[\prod_{j=0}^{n-1}(\alpha_j+\alpha_{j+1})\right](\alpha_0 \pm ik)} \tag{4.27a}$$

Again this last series is absolutely and uniformly convergent for all x such that $\mathrm{Re}x > x_0 = \frac{\log A}{m}$. The series for $\rho_{\pm}(k,\lambda)$ are finite for $m < \lambda < M$, and given by

$$\rho_{\pm}(k,\lambda)=\sum_{n=0}^{N(M)}(\frac{-1}{\pi})^{n+1}\int_{\frac{m}{2}}^{\infty}d\alpha_0\cdots\int_{\frac{m}{2}}^{\infty}d\alpha_n\frac{(\prod_{j=0}^{n}\nu(\alpha_j))\delta(\lambda-2\sum_{j=0}^{n}\alpha_j)}{[\prod_{j=0}^{n-1}(\alpha_j+\alpha_{j+1})](\alpha_0\pm ik)} \tag{4.27b}$$

In the appendix we shall check directly that for $x > x_0$, eqn. (4.27) gives a solution of the Schrödinger equation with the potential given by (4.24) for arbitrary $\nu(\alpha)$ with A finite.

Next we have to extend our results to the region $0 < x \leq x_0$. It is clear from (4.24) and (4.27) that we have to deal with an operator, $K(x)$, which depends on a parameter, x, where we take Re$x > 0$, and $K(x)$ operates on functions $\phi(\beta)$, $\frac{m}{2} \leq \beta < \infty$, where $\phi \in L_2(\frac{m}{2},\infty)$. We define K as:

$$[K(x).\phi](\alpha) \equiv \int_{\frac{m}{2}}^{\infty} K(\alpha,\beta;x)\phi(\beta)d\beta \ , \tag{4.28}$$

where

$$K(\alpha,\beta;x) = (-\frac{1}{\pi})\frac{\nu(\beta)e^{-2\beta x}}{[\alpha+\beta]} \ , \quad Rex \geq 0. \tag{4.29}$$

The first thing to notice from (4.3) and (4.22) is that for ν which is the discontinuity of a Yukawian $S(k)$, $\|K\|$ the $L_2(\frac{m}{2},\infty)$ norm of K is bounded,

$$\|K\| \leq Ae^{-mx}. \tag{4.30}$$

All we need now is contained in the simpler integral equation

$$W(\alpha,x) = 1 + \int_{\frac{m}{2}}^{\infty} K(\alpha,\beta;x)W(\beta,x)d\beta \ , \quad \text{Re}x \geq 0, \tag{4.31}$$

with K given by (4.29) and x a parameter. For $x \equiv 0$ Eq. (4.31) is just the Noyes-Wong[13] equation derived from the N/D method. The integral equation (4.31) is of the Fredholm type and the solution $W(\alpha,x)$ exists. For $x > x_0 = \frac{\log A}{m}$, (4.30) guarantees that $\|K\|$ is less than unity and hence

$$W(\alpha,x) = 1 + \sum_{n=1}^{\infty}\int_{\frac{m}{2}}^{\infty} K^{(n)}(\alpha,\beta;x)d\beta \ , \quad \text{Re}x > x_0, \tag{4.32}$$

and the series is absolutely convergent.We also have:

$$V(x) = -2\frac{d}{dx}[\frac{-1}{\pi}\int_{\frac{m}{2}}^{\infty}\nu(\alpha)e^{-2\alpha x}W(\alpha,x)d\alpha] \ , \quad \text{Re}x > x_0. \tag{4.33}$$

By substituting (4.32) in (4.33) for W we obtain the same series for $V(x)$ as in (4.24). Similarly, one can show that,

$$h_{\pm}(k,x) = \frac{-1}{\pi}\int_{\frac{m}{2}}^{\infty}\frac{\nu(\alpha)e^{-2\alpha x}}{(\alpha \pm ik)}W(\alpha,x)d\alpha \ , \quad \text{Re}x > x_0, \tag{4.34}$$

gives the identical result to (4.27a).

The relation between $V(x)$ and the kernel K can be further simplified by using the following relation,

$$\log\left[\mathrm{Det}(1 - gK(x))\right] = -\mathrm{Tr}\sum_{n=1}^{\infty}\frac{g^n}{n}K^{(n)}(x) \;, \tag{4.35}$$

which is valid whenever the series on the right hand side is absolutely convergent. For $x > x_0$, we have absolute convergence in our case and we can set $g = 1$,

$$\log(\mathrm{Det}(1 - K(x))) = -\sum_{n=1}^{\infty}\left(\frac{-1}{\pi}\right)^n\int_{\frac{m}{2}}^{\infty}d\alpha_0\cdots\int_{\frac{m}{2}}^{\infty}d\alpha_{n-1}\frac{\left(\prod_{j=0}^{n-1}\nu(\alpha_j)e^{-2\alpha_j x}\right)}{\left[\prod_{j=0}^{n-1}(\alpha_j + \alpha_{j+1})\right]n} \tag{4.36}$$

where $\alpha_n \equiv \alpha_0$. Differentiating with respect to x,

$$\frac{d}{dx}\log[\mathrm{Det}(1 - K(x))] =$$

$$\sum_{n=1}^{\infty}\left(\frac{-1}{\pi}\right)^n\int_{\frac{m}{2}}^{\infty}d\alpha_0\cdots\int_{\frac{m}{2}}^{\infty}d\alpha_{n-1}\frac{\left(\prod_{j=0}^{n-1}\nu(\alpha_j)e^{-2\alpha_j x}\right)}{\left[\prod_{j=0}^{n-1}(\alpha_j + \alpha_{j+1})\right]}\left(\frac{2\sum_{j=0}^{n-1}\alpha_j}{n}\right). \tag{4.37}$$

We can write $2\sum_{j=0}^{n-1}\alpha_j = (\alpha_0+\alpha_1)+(\alpha_1+\alpha_2)+\cdots+(\alpha_{n-1}+\alpha_0)$, and then we have n integrals each with one less factor in the product that appears in the denominator. These integrals are equal by symmetry and thus the factor n is cancelled and we have

$$\frac{d}{dx}\log[\mathrm{Det}(1 - K(x))]$$

$$= \sum_{n=0}^{\infty}\left(\frac{-1}{\pi}\right)^{n+1}\int_{\frac{m}{2}}^{\infty}d\alpha_0\cdots\int_{\frac{m}{2}}^{\infty}d\alpha_{n-1}\frac{\left(\prod_{j=0}^{n-1}\nu(\alpha_j)\epsilon^{-2\alpha_j x}\right)}{\left[\prod_{j=0}^{n-2}(\alpha_j + \alpha_{j+1})\right]}. \tag{4.38}$$

Comparing this with (4.24) we obtain Dyson's formula[4]

$$V(x) = -2\frac{d^2}{dx^2}\log[\mathrm{Det}(1 - K(x))] \tag{4.39}$$

The series in (4.38) is absolutely convergent only for $x_0 < x$. When $A > 1$, we need to continue the expression in (4.39) for the whole interval $0 \le x < \infty$. There are two separate questions to consider. We shall show below that the Fredholm determinant, $\mathrm{Det}(1 - K(x))$, will not vanish for the whole interval $0 \le x < \infty$, if $\nu(\alpha)$ is the discontinuity of a bonafide, $S(k)$. The second problem is to write a convergent series for $\mathrm{Det}(1 - K(x))$ for any $0 < A < \infty$, and $0 \le x < \infty$.This we can do since the determinants in the Fredholm formula for $K(\alpha,\beta;x)$ can be explicitly evaluated. The standard result gives

$$\mathrm{Det}(1-K(x)) =$$

$$1+\sum_{n=1}^{\infty}\frac{1}{n!}\left(\frac{-1}{\pi}\right)^n\int_{\frac{m}{2}}^{\infty}d\alpha_0\cdots\int_{\frac{m}{2}}^{\infty}d\alpha_n\left(\prod_{j=0}^{n-1}\nu(\alpha_j)e^{-2\alpha_j x}\right)\mathrm{Det}_n\left|\frac{1}{\alpha_i+\alpha_j}\right| \tag{4.40}$$

where

$$\mathrm{Det}_n\left|\frac{1}{\alpha_i+\alpha_j}\right| \equiv \mathrm{Det}\begin{pmatrix} \frac{1}{2\alpha_1} & \frac{1}{\alpha_1+\alpha_2} & \cdots & \frac{1}{\alpha_1+\alpha_n} \\ \vdots & & & \\ \frac{1}{\alpha_n+\alpha_1} & \cdots & \cdots & \frac{1}{2\alpha_n} \end{pmatrix}. \tag{4.41}$$

The series in (4.40) is absolutely convergent for all x such that $\mathrm{Re}x \geq 0$, as long as A, given in (4.22), is finite. In addition the determinant (4.41) can be evaluated[14],

$$\mathrm{Det}_n\left|\frac{1}{\alpha_i+\alpha_j}\right| = \frac{1}{2^n}\frac{1}{\left(\prod_{j=1}^{n}\alpha_j\right)}\left[\prod_{i<j}^{n}\left(\frac{(\alpha_i-\alpha_j)^2}{(\alpha_i+\alpha_j)^2}\right)\right]. \tag{4.42}$$

Substituting (4.42) in (4.40), we get an expression for $V(x)$ valid for all x, $0 \leq x < \infty$ which gives V in terms of the discontinuity of $S(k)$, $\nu(\alpha)$:

$$V(x) = -2\tfrac{d^2}{dx^2}\log$$

$$\left[1+\sum_{n=1}^{\infty}\frac{(-1)^n}{n!}\int_{\frac{m}{2}}^{\infty}d\alpha_1\cdots\int_{\frac{m}{2}}^{\infty}d\alpha_n\left(\prod_{j=1}^{n}\frac{\nu(\alpha_j)e^{-2\alpha_j x}}{2\pi\alpha_j}\right)\prod_{i<j}^{n}\left(\frac{(\alpha_i-\alpha_j)^2}{(\alpha_i+\alpha_j)^2}\right)\right]. \tag{4.43}$$

Finally, we prove that when $\nu(\alpha)$ is given to us as the discontinuity of a physical $S(k)$, then $\mathrm{Det}(1-K(x))$ cannot vanish for $0 \leq x < \infty$. Since $K(x)$ is Fredholm, the Fredholm alternative holds, and $\mathrm{Det}(1-K(x)) = 0$ if and only if the homogeneous equation,

$$W(\alpha;x) = -\frac{1}{\pi}\int_{\frac{m}{2}}^{\infty}\frac{\nu(\beta)e^{-2\beta x}}{\alpha+\beta}W(\beta;x)d\beta\ , \tag{4.44}$$

has a solution. This cannot happen because it would imply the existence of a solution of the homogeneous Marchenko equation, which we know has no solutions for any $S(k)$ satisfying (a), (b), (c). Indeed multiplying both sides of (4.44) by $-\frac{\nu(\alpha)}{\pi}e^{-\alpha x}$, and defining $Z(\alpha;x)$ as

$$Z(\alpha;x) \equiv -\frac{\nu(\alpha)}{\pi}e^{-\alpha x}W(\alpha;x)\ , \tag{4.45}$$

we get for Z,

$$Z(\alpha;x) = -\frac{\nu(\alpha)}{\pi}e^{-\alpha x}\int_{\frac{m}{2}}^{\infty}\frac{e^{-\beta x}Z(\beta;x)}{(\alpha+\beta)}d\beta\,. \tag{4.46}$$

Letting $U(y;x)$, be

$$U(y;x) \equiv \int_{\frac{m}{2}}^{\infty} Z(\alpha;x)e^{-\alpha y}d\alpha\,, \tag{4.47}$$

and using the identity

$$\frac{e^{-(\alpha+\beta)x}}{(\alpha+\beta)} = \int_x^{\infty} e^{-(\alpha+\beta)z}dz\,,$$

we obtain for $U(y;x)$

$$U(y;x) = \int_x^{\infty} F(y+z)U(z;x)dz\,. \tag{4.49}$$

This is just the homogeneous Marchenko equation and is known not to have solutions for any x such that $0 \le x < \infty$. The proof is given in the text book of De Alfaro and Regge[15] and depends essentially on the fact that $|S(k)| = 1$ for real k.

Before closing this section we should remark about the analyticity of $V(x)$ for Re$x > 0$. From references 7 and 8 it is known that when $V(x)$ has the representation, $V(x) = \int_m^{\infty} C(\lambda)e^{-\lambda x}d\lambda$, the resulting $S(k)$ will be of the Yukawian class, i.e. will have the properties (d) and (e). The converse has never been proved. Namely, given an $S(k)$ which is Yukawian, is the potential obtained by inverse scattering guaranteed to have a representation $V(x) = \int_m^{\infty} C(\lambda)e^{-\lambda x}d\lambda$, and hence be analytic for $Re x > 0$. It is already clear from (4.43) that the Fredholm determinant, $\mathrm{Det}(1-K(x))$, is analytic for Re$x \ge 0$ and has a Laplace transform representation. The only difficulty then could come from the zeros of the determinant in the half-plane, Re$x >$ 0. We have already excluded these zeros on the real axis, $0 \le x < \infty$.

The expression (4.27) for the Jost solutions $h_{\pm}(k,x)$ is only valid for $x > (\log A/m)$. To obtain an expression valid for all Re$x \ge 0$ we have to first write the Fredholm series solution for $W(\alpha,x)$ and then use the result in (4.34) to get $h_{\pm}(k,x)$ for all $Re \ge 0$. The solution $W(\alpha,x)$ exists in the half-line, $x \ge 0$, since $\mathrm{Det}(1-K(x))$ does not vanish there.

V. Potentials for Arbitrary $\nu(\alpha)$

In the preceding discussion the functions $\nu(\alpha)$ were not arbitrary but were obtained from the discontinuity of a physical Yukawian $S(k)$. Such an $S(k)$ for example has no singularities in k except on the pure imaginary axis, where also the poles of the point spectrum happen to be. This information is somehow contained in $\nu(\alpha)$ but not in a transparent way.

The question that arises is: if we are given an arbitrary summable, $\nu(\alpha)$, $\frac{m}{2} \leq \alpha < \infty$, such that

$$\frac{1}{\pi}\int_{\frac{m}{2}}^{\infty} \frac{|\nu(\alpha)|}{\alpha} d\alpha = A < \infty , \tag{5.1}$$

what conditions do we need to impose on $\nu(\alpha)$ to guarantee that the potential, $V(x)$, obtained from $\nu(\alpha)$ via the formula (4.43) will have a non-vanishing determinant and an $S(k)$ whose discontinuity is given by $\nu(\alpha)$.

It is obvious from the discussion of the previous section that the only fact we need to establish is the non-vanishing of $\mathrm{Det}(1-K(x))$ for the given $\nu(\alpha)$ in the interval $0 \leq x < \infty$.

First, one can give a sufficient condition. We have from (4.43),

$\Delta(x) \equiv \mathrm{Det}(1-K(x)) =$

$$1+\sum_{n=1}^{\infty}\frac{1}{n!}\int_{\frac{m}{2}}^{\infty}\cdots\int_{\frac{m}{2}}^{\infty}\left(\prod_{j=1}^{n}\frac{-\nu(\alpha_j)e^{-2\alpha_j x}d\alpha_j}{2\pi\alpha_j}\right)\prod_{i<j}^{n}\left(\frac{(\alpha_i-\alpha_j)^2}{(\alpha_i+\alpha_j)^2}\right). \tag{5.2}$$

Using (5.1), it follows now that

$$|\Delta(x)-1| \leq |e^{A}-1|, \quad 0 \leq \mathrm{Re}x < \infty. \tag{5.3}$$

This inequality shows that if

$$A < 2 \log 2, \tag{5.4}$$

the Fredholm determinant, $\Delta(x)$, will not have any zeros in the half plane $\mathrm{Re}x \geq 0$. the inequality (5.4) thus gives us a sufficient condition for the existence of a solution of the equation (4.31), $W(\alpha, x)$ for all $\mathrm{Re}x \geq 0$.

The S-matrix, is given by

$$S(k) = \frac{h_+(k,0)}{h_-(k,0)} = \frac{1+\frac{1}{\pi}\int_{\frac{m}{2}}^{\infty}\frac{w(\alpha)}{\alpha+ik}d\alpha}{1+\frac{1}{\pi}\int_{\frac{m}{2}}^{\infty}\frac{w(\alpha)}{\alpha-ik}d\alpha}, \tag{5.5}$$

with

$$w(\alpha) \equiv -\nu(\alpha)W(\alpha,0), \tag{5.6}$$

as is evident from equation (4.34), which now can be taken to be valid for all x with $Rex \geq 0$.

Equation (4.31) for $W(\alpha, x)$ becomes the Noyes-Wong[13] equation for $x = 0$ and we get

$$W(\alpha,0) = 1 + \frac{(-1)}{\pi}\int_{\frac{m}{2}}^{\infty}\frac{\nu(\beta)W(\beta,0)}{\alpha+\beta}d\beta, \tag{5.7}$$

where we note that $W(\alpha,0) = h_-(i\alpha)$, $\alpha > 0$. It follows from (2.5) and (5.5)

$$\nu(\alpha) = \frac{1}{2i}[S(i\alpha + 0) - S(i\alpha - 0)] = \frac{w(\alpha)}{[1 + \frac{1}{\pi}\int_{\frac{m}{2}}^{\infty} \frac{w(\beta)}{\alpha+\beta} d\beta]}; \tag{5.8}$$

Using (5.6) and (5.7) we see that (5.8) is an identity guaranteeing that $\nu(\alpha)$ is the discontinuity of $S(k)$.

The condition (5.4) on $\nu(\alpha)$ is just a sufficient condition to guarantee that ν has the needed properties. One would like a necessary and sufficient condition on $\nu(\alpha)$ to assure that $\Delta(x)$ will have no zeros in the half-plane, $\mathrm{Re}x \geq 0$. This can be done, however the condition is not simple and consequently not very useful. We consider functions $\phi(\alpha) \in L_2(\frac{m}{2}, \infty)$, and the condition will be

$$\min_{\phi(\alpha)\in L_2(\frac{m}{2},\infty),\mathrm{Re}x>0} \left| \int_{\frac{m}{2}}^{\infty} d\alpha \int_{\frac{m}{2}}^{\infty} d\beta \, \phi(\alpha)[\delta(\alpha - \beta) - \frac{\nu(\beta)e^{-2\beta x}}{\pi(\alpha + \beta)}]\phi(\beta) \right| > 0. \tag{5.9}$$

This will guarantee that the homogeneous integral equation (4.44) has no solutions for $Rex \geq 0$, and serve as a necessary and sufficient condition for the existence of a $V(x)$ whose $S(k)$ has $\nu(\alpha)$ as its discontinuity along the cut.

We close by stating the following theorem.

Theorem :

Given any summable function $\nu(\alpha)$ satisfying (5.1), (5.4), and\or (5.9), then $V(x)$ given by

$$V(x) = -2\tfrac{d^2}{dx^2} \log$$

$$\left[1 + \sum_{n=1}^{\infty} \frac{1}{n!} \int_{\frac{m}{2}}^{\infty} \cdots \int_{\frac{m}{2}}^{\infty} \left(\prod_{j=1}^{n} \frac{-\nu(\alpha_j)e^{-2\alpha_j x} d\alpha_j}{2\pi\alpha_j} \right) \prod_{i<j}^{n} \left(\frac{(\alpha_i - \alpha_j)^2}{(\alpha_i + \alpha_j)^2} \right) \right].$$

will, when used as a potential in the Schrödinger equation, lead to a Yukawian class $S(k)$. The discontinuity of $S(k)$ along the cut $k = i\alpha$, $\frac{m}{2} \leq \alpha < 0$ will be $\nu(\alpha)$. In addition $\nu(\alpha)$ determines $S(k)$ and the discrete spectrum uniquely.

The proof of this theorem is evident from the arguments given in the last two sections.

Appendix

The Schrödinger equation for the functions,$h_\pm(k,x)$, defined in (3.5) is:

$$\frac{d^2h_\pm}{dx^2}(k,x) \mp 2ik\frac{dh_\pm}{dx}(k,x) = V(x)h_\pm(k,x). \qquad (A-1)$$

Using the Laplace representations for $h_\pm$ and V:

$$V(x) = \int_m^\infty C(\lambda)e^{-\lambda x}d\lambda,$$

$$h_\pm(k,x) = 1 + \int_m^\infty \rho_\pm(k,\lambda)e^{-\lambda x}d\lambda, \qquad (A-2)$$

one gets the Martin integral equation:

$$\lambda(\lambda \pm 2ik)\rho_\pm(k,\lambda) = C(\lambda) + \theta(\lambda-2m)\int_m^{\lambda-m} C(\lambda-\gamma)\rho_\pm(k,\gamma)d\gamma. \quad (A-3)$$

This last equation for $\rho_\pm$ is equivalent to the Schrödinger equation when the integrals in (A.2) are convergent for all Re$x > 0$.

Our solutions for $C(\lambda)$ and $\rho_\pm(k,\lambda)$ in terms of $\nu(\alpha)$, given in (4.21) and (4.27b), are

$$C(\lambda) = 2\lambda\sum_{n=0}^{N(M)}\left(\frac{-1}{\pi}\right)^{n+1}\int_{\frac{m}{2}}^\infty d\alpha_0\cdots\int_{\frac{m}{2}}^\infty d\alpha_n\frac{\left(\prod_{j=0}^n\nu(\alpha_j)\right)}{\left(\prod_{j=0}^{n-1}(\alpha_j+\alpha_{j+1})\right)}\delta(\lambda-2\sum_{j=0}^n\alpha_j) \qquad (A-4)$$

and

$$\rho_\pm(k,\lambda) = \sum_{n=0}^{N(M)}\left(\frac{-1}{\pi}\right)^{n+1}\int_{\frac{m}{2}}^\infty d\alpha_0\cdots\int_{\frac{m}{2}}^\infty d\alpha_n\frac{\left(\prod_{j=0}^n\nu(\alpha_j)\right)\delta(\lambda-2\sum_{j=0}^n\alpha_j)}{\left[\prod_{j=0}^{n-1}(\alpha_j+\alpha_{j+1})\right](\alpha_0\pm ik)} \qquad (A-5)$$

These are valid for $m \leq \lambda < M$, for any finite M, and $N(M)$ is an integer defined in (4.18). We check directly that (A-4) and (A-5) gives a solution of (A-3). We calculate the integral in (A-3)

$$\int_m^{\lambda-m} C(\lambda-\gamma)\rho_\pm(k,\gamma)d\gamma = \int_m^{\lambda-\gamma} 2(\lambda-\gamma)\Big\{\sum_{n=0}^N\left(\tfrac{-1}{\pi}\right)^{n+1}\times$$

$$\times\int_{\frac{m}{2}}^\infty d\alpha_0\cdots\int_{\frac{m}{2}}^\infty d\alpha_n\frac{\prod_{j=0}^n\nu(\alpha_j)}{\prod_{j=0}^{n-1}(\alpha_j+\alpha_{j+1})}\delta(\lambda-\gamma-2\sum_{j=0}^n\alpha_j)\Big\}$$

$$\times\Big\{\sum_{l=0}^N\left(\frac{-1}{\pi}\right)^{l+1}\int_{\frac{m}{2}}^\infty d\beta_0\cdots\int_{\frac{m}{2}}^\infty d\beta_l\frac{\prod_{i=0}^l\nu(\beta_i)}{\prod_{i=0}^{l-1}(\beta_i+\beta_{i+1})}\frac{\delta(\gamma-2\sum_{i=0}^l\beta_i)}{(\beta_l\pm ik)}\Big\}d\gamma$$

We set $(n+l+1) \equiv r$, relabel the β's as $\beta_0 = \alpha_{n+1}, \beta_1 = \alpha_{n+2}, \cdots, \beta_l = \alpha_r$, and integrate over γ to get

$$\int_m^{\lambda-m} C(\lambda-\gamma)\rho_\pm(k,\gamma)d\gamma =$$

$$4\sum_{r=1}^{\infty}\left(\frac{-1}{\pi}\right)^{r+1}\int_{\frac{m}{2}}^{\infty} d\alpha_0 \cdots \int_{\frac{m}{2}}^{\infty} d\alpha_r \frac{\prod_{j=0}^{r}\nu(\alpha_j)}{\prod_{j=0}^{r-1}(\alpha_j+\alpha_{j+1})}$$

$$\times \frac{\delta(\lambda - 2\sum_{j=0}^{r}\alpha_j)}{(\alpha_r \pm ik)}\left[\sum_{n=0}^{r-1}\{(\alpha_n+\alpha_{n+1})(\sum_{j=0}^{n}\alpha_j)\}\right]. \quad (A-6)$$

The last bracket satisfies the identity

$$\left[\sum_{n=0}^{r-1}\{(\alpha_n+\alpha_{n+1})(\sum_{j=0}^{n}\alpha_j)\}\right] = \left(\sum_{j=0}^{r}\alpha_j\right)^2 - \alpha_r\left(\sum_{j=0}^{r}\alpha_j\right). \quad (A-7)$$

Substituting this result above and adding $\pm ik \mp ik$ in the numerator we get

$$\int_m^{\lambda-m} C(\lambda-\gamma)\rho_\pm(k,\gamma)d\gamma = \lambda(\lambda \pm 2ik)\rho_\pm - C(\lambda), \quad (A-8)$$

which is the desired result. One can also check directly that for $x > x_0$, the convergent series for $h_\pm$ and $V(x)$ given by (4.27a) and (4.24) do satisfy the differential equation (A-1).

References and Footnotes

[1] V. Bargmann, Phys. Rev. **75**, 301 (1949); and Rev. Mod. Phys. **21**, 488 (1949).

[2] I.M. Gelfand and B.M. Levitan, Izv. Akad. Nauk. SSSR, Ser. Mat. **15**, 309 (1951), translation in Am. Math. Soc. Translation (2), **1**, 253 (1950) . Also V. A. Marchenko, Dokl. Akad. Nauk. SSSR **72**, 457(1950); and **104**, 695(1955).

[3] L.D. Faddeev, Usp. Mat. Nauk **14**, 57 (1959); translated in J. Math. Phys. **4**, 72 (1963).

[4] F.J. Dyson, in "*Studies in Mathematical Physics*"; E. Lieb, B. Simon and A. S. Wightman, editors; (Princeton Univ. Press, Princeton, N.J., 1976), pages 151-167.

[5] See for example K. Chadan and P.C. Sabatier, "*Inverse Problems in Scattering Theory*", 2nd ed.(Springer-Verlag, 1989).

[6] A. Martin, Nuovo Cim. **19**, 1257 (1961).

[7] A. Martin, Nuovo Cim. **14**, 403 (1959).

[8] R. Blankenbecler, M.L. Goldberger, N.N. Khuri, and S.B. Treiman, Annals of Phys. **10**, 62 (1960).

[9] D.J. Gross and B.J. Kayser, Phys. Rev.**152**, 1441 (1966).

[10] H. Cornille,J.M.P. **8**,2268 (1967), and J.M.P. **11**, 61 (1970).

[11] N.N. Khuri, Ann. of Phys. **201**, (1990).

[12] This condition can be weakened and it is enough to assume that $|S(k) - 1| \to 0$ as $|k| \to \infty$ uniformly for all k with $\mathrm{Im}k \geq 0$.

[13] H.P. Noyes and D.Y. Wong, Phys. Rev Lett. **3**, 191 (1959).

[14] V. DeAlfaro and T. Regge, Nuovo Cim. **20**, 956 (1966).

[15] V. DeAlfaro and T. Regge, "*Potential Scattering*", (John Wiley and Sons, Inc., New York, (1965), page 157.

The Inverse of an Inverse Problem

J. Leon and P.C. Sabatier

Département de Physique Mathématique, Université Montpellier II, F-34095 Montpellier Cedex 05, France

Meaning a problem as an "inverse" problem makes sense only if the "direct" problem is a general problem of physical sciences, and is easier to solve than the "inverse" one. In "direct scattering problems" we introduce the shape of a target, or that of an interaction, and we derive asymptotic results, to be compared to experimental results. Going backward is a difficult problem, which depends specifically on each individual case, which may be ill-posed and is certainly more complicated than going forward. For this reason, we call "direct problems" the forward ones, and "inverse problems" the backward ones. But what happens when the backward problem becomes simpler than the forward problem ?

Let us first observe that in almost all known studies, the analysis of backward problems is *not* what it should be for an inverse problem. Indeed, assume a model is described by giving a mapping M from a set C of parameters into a set E of results (where E contains all possible calculated and measured results together). A correct analysis of the inverse problem E to C must be done by allowing C to contain all admissible physical parameters and disentangling completely all possible bijections between subsets of C and subsets of E. However, this is never done. In most studies, authors select a special class of parameters, for which the direct and inverse problems are easy to do, e.g. the separable case in the quantum scattering theory with non local interactions. In this classical example, the special choice of C makes it possible to derive the solvability conditions and to construct the solution the key condition being the generalized Levinson's theorem (André Martin, 1958). In a narrower class of studies, C is chosen for physical reasons, but only partial results or instable reconstructions, are possible. A classical example is the inverse problem at fixed energy with interactions that are superpositions of Yukawa potentials, with the classical uniqueness theorem and formal reconstruction procedures of Martin and Targonsky (1961). A classical example for analysis going as far as possible to the complete one, which was defined above, and applying to a physical problem, is that of the inverse problem : cross section $\longrightarrow$ scattering amplitude (André Martin 1973) but of course, one cannot expect to obtain for physical problems a perfect analysis as the Gelfand Levitan determination of a differential operator from its spectral function, or as similar results obtained for the Schrödinger inverse problem on the line or the Zakharov Shabat inverse problem (see e.g. Chadan-Sabatier 1989, Ch. XVII).

Speaking of mathematical inverse problems, we are led to consider the various inverse scattering problems studied only for solving non linear evolution equations. This research completely renewed our vision of classical inverse problems. In the seventies, it became clear that the key to inverse methods is a way of associating two trajectories in C and E, the first one being interesting, and described by a non linear evolution equation, the second one being described trivially. In the same way, Bäcklund and Darboux transformations in C, and their images in E, define joint routes for exploring these spaces. In turn the idea of a joint exploration of C and E yields new ways of handling inverse problems and of dealing not only with reconstructions but also with ill-posedness questions (P.C. Sabatier 1987). Hence, one steadily progressed towards a symmetric way of dealing with C and E, where no problem can be considered the *direct one*, or the *inverse one*. Still further a recent inverse method goes beyond the full symmetry and makes the usual direct scattering problem an inverse of the inverse scattering problem. It is the so called method of "singular dispersion laws" (Léon 1990) which achieves the most elaborate combination between algebraic methods and $\bar{\partial}$-methods. We tersely describe it.

A couple (Ω, Λ) of 2x2-matrix-valued functions of k, x, t is defined in a space that we call E. The solution $\psi(k,x,t)$ of the $\bar{\partial}$-problem

$$\frac{\partial}{\partial \bar{k}}\psi(k,x,t) = \psi(k,x,t)R(k,x,t), \qquad k \in D \subset C, \tag{1}$$

where $\psi(k)$ is invertible and given on ∂D, depends on Ω, Λ (the dispersion laws) through the dispersion equations

$$\begin{aligned} \partial_t R &= [R, \Omega] \\ \partial_x R &= [R, \Lambda] \end{aligned} \tag{2}$$

The function ψ is used to define a couple U, V of 2x2-matrix-valued functions of k, x, t by means of the formulas

$$\begin{aligned} U\psi &= \partial_x \psi - \psi\Lambda \\ V\psi &= \partial_t \psi - \psi\Omega \end{aligned} \tag{3}$$

It is easy to derive from (1), (2), (3) the analytic structure of U and V :

$$\frac{\partial U}{\partial \bar{k}} = -\psi \frac{\partial \Lambda}{\partial \bar{k}} \psi^{-1} \tag{4.a}$$

$$\frac{\partial V}{\partial \bar{k}} = -\psi \frac{\partial \Omega}{\partial \bar{k}} \psi^{-1} \tag{4.b}$$

and, after a few trivial calculations, one proves that the two following evolution equations (resp. in E or in C) are equivalent (i.e. images of each other)

$$\frac{\partial \Lambda}{\partial t} - \frac{\partial \Omega}{\partial x} + [\Omega, \Lambda] = 0 \tag{5}$$

$$\frac{\partial U}{\partial t} - \frac{\partial V}{\partial x} + [U, V] = 0 \tag{6}$$

At this point, the problem has been described in such a way that the input is the set $\{R, \Lambda, \Omega\}$ together with the boundary condition for ψ on ∂D and the condition (5). Then, any solution $\{\psi, U, V\}$ of the $\bar{\partial}$-equations (1) and (4) yields a solution of the nonlinear evolution equation (6).

This way of processing takes as input the "spectral data" $R(k, x, t)$ and when the $\bar{\partial}$-problem (1) is equivalent to a scattering problem, the inversion of our procedures reduce to the direct scattering problem (determination of $\{R, \Lambda, \Omega\}$ from $\{\psi, U, V\}$ obeying (6) and (3)), the solution of which is not in general explicitely feasible.

The simplest example is the famous nonlinear Schrödinger equation (NLS) obtained for $D = C$, $\psi \to 1$, for $k \to \infty, \Lambda = ik\sigma_3, \Omega = 2ik^2\sigma_3$ (they automatically verify (5)). The solution of (4) gives in particular

$$U(k, x, t) = -ik\sigma_3 + Q(x, t), \quad diag\{Q\} = 0 \tag{7}$$

and (6) becomes

$$Q_t = i\sigma_3 Q_{xx} - 2i\sigma_3 Q^3 \tag{8}$$

which reduces to the NLS eq. in the reduction $Q\sigma_2 = \sigma_2 Q^*$. One step further is achieved by assuming that Ω nonanalytic part, f.i :

$$\Omega = 2ik^2\sigma_3 + \iint_C \frac{d\lambda \wedge d\bar{\lambda}}{\lambda - k} g(\lambda)\sigma_3 \tag{9}$$

Then the integrable evolution (6) becomes the following coupled system

$$\begin{aligned} Q_t - i\sigma_3 Q_{xx} + 2i\sigma_3 Q^3 &= i[\sigma_3, \textstyle\iint_C g\psi\sigma_3\psi^{-1}] \\ \psi_x + ik[\sigma_3, \psi] &= Q\psi \end{aligned} \tag{10}$$

which is relevant in problems of laser-plasma interaction (ψ is related to the electrostatic field and Q to the electronic density).

This "algebraic-analytic" approach is very rich for it furnishes the largest class of integrable nonlinear evolution equations obtained when neither Ω nor Λ are analytic in k (Léon 1990).

The present approach also works for two-dimensional problems when replacing (1) and (2) with (Boiti et al 1988)

$$\frac{\partial}{\partial \bar{k}}\psi = \iint_C d\lambda \wedge d\bar{\lambda}\psi(\lambda)R(k,\lambda) \tag{11}$$

$$\partial_t R(k,\lambda) = R(k,\lambda)\Omega(\lambda) - \Omega(k)R(k,\lambda)$$

$$\partial_x R(k,\lambda) = R(k,\lambda)\Lambda(\lambda) - \Lambda(k)R(k,\lambda) \tag{12}$$

$$\partial_y R(k,\lambda) = R(k,\lambda)\Gamma(\lambda) - \Gamma(k)R(k,\lambda).$$

REFERENCES

BOITI M., LEON J., MARTINA L., PEMPINELLI F.: Integrable nonlinear evolutions in 2+1 dimensions with nonanalytic dispersion relations. J. Phys. A**21**, 3611 (1988)

CHADAN K., SABATIER P.C.: "Inverse Problems in Quantum Scattering Theory" 2nd Edition, revised and expanded, Springer-Verlag (New-York, Heidelberg, London, Paris, Tokyo) 1989

LEON J.: Nonlinear evolutions with singular dispersion laws and forced systems. Phys. Lett. A **144**, 444 (1990)

MARTIN A. (1958): On the validity of Levinson's theorem for non-local interactions, Nuovo Cimento **7**, 607-627

MARTIN A. and TARGONSKY G.Y. (1961): On the uniqueness of a potential fitting a scattering amplitude at a given energy, Nuovo Cimento **20**, 1182-1190

MARTIN A.: Reconstruction of scattering amplitude from differential cross sections, Lectures given at the Adriatic meeting on particle physics in Rovjni (1973)

SABATIER P.C.: A few geometrical features on inverse and ill-posed problems in "Inverse and Ill-posed problems" Engl. and Groetsch eds. (Ac. Press, New-York 1987)

Weak-Star Compactness and Its Relevance to Analyticity Problems

M. Ciulli[1], *S. Ciulli*[1], *and T.D. Spearman*[2]

[1]Laboratoire de Physique Mathématique, Université Montpellier II, F-34095 Montpellier Cedex 05, France
[2]School of Mathematics, Trinity College, Dublin, Ireland

Abstract

Analytic continuation is inherently unstable; to rectify this situation it is necessary to introduce appropriate ancillary information. This paper is concerned with the use of bounds for this purpose. Stabilization is normally achieved by a restriction to compact sets within the function space, whereas bounds are expressed in terms of balls $\|X\| \leq \delta$ which are not compact in the norm but only in the weak-* sense. The implications of this are examined and it is shown how and in what circumstances stabilization can be attained.

This work is dedicated to André Martin who has not only exercised a great influence generally on the uses of analyticity in modern physics, but has also derived many of the rigorous bounds whose important applications include that described in this paper.

1. Introduction

Consider the problem of analytic continuation from a segment γ lying in the interior of the holomorphy domain of a scattering amplitude $X(z)$. We shall suppose that $X(z)$ is real-analytic, defined in a plane with cuts extending from branch-points on the real axis, and that the finite closed segment γ lies on the real axis but does not overlap with the cuts. A standard transformation allows the cuts to be mapped on to the unit circle which thus becomes the boundary of the domain of holomorphy.

A function $X(z)$ which is of real type (i.e. $X(z) = X^*(z^*)$) and is holomorphic in the unit disk may be represented by a Poisson (Schwarz-Villat) integral of the form

$$X(z) = \int_0^{2\pi} P(z,\theta)x(\theta)d\theta \tag{1}$$

where the boundary function $x(\theta)$ is $x(\theta) \equiv ReX(e^{i\theta})$ and the (complexified) Poisson kernel $P(z,\theta)$ is defined as

$$P(z,\theta) = \frac{1}{2\pi}\frac{e^{i\theta}+z}{e^{i\theta}-z}. \tag{2}$$

We shall denote the restriction of the holomorphic function $X(z)$ to the segment γ by $X^\gamma(z)$. Since any holomorphic function $X(z)$ is specified by

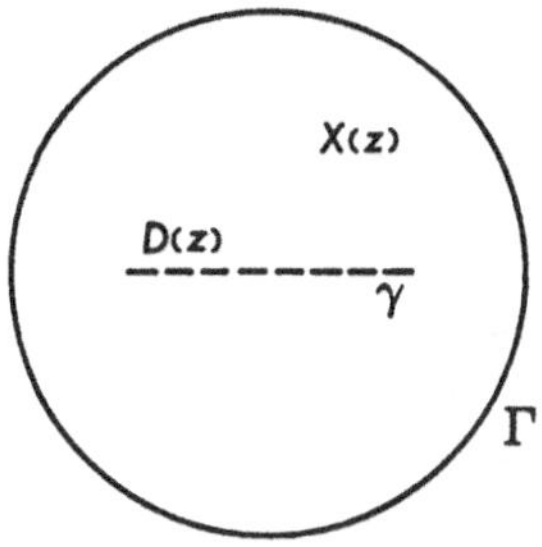

Figure 1: *The real data function $D(z)$, given with errors on the segment γ, will be taken to be continuous but need not be differentiable and thus may not be the restriction on γ of any harmonic function. The problem of interest is to approximate to $D(z)$ by means of functions $X(z)$ which are holomorphic in the unit disk.*

the (real) boundary function $x(\theta)$ we may express the continuation problem schematically as follows

$$D(z) \longrightarrow X_0^\gamma(z) \longrightarrow x_0(\theta). \tag{3}$$

A key element in the definition of the problem is to specify a topology on the space of functions $D(z), X^\gamma(z)$ defined on γ. For example a standard χ^2 measure would be expressed in terms of an L^2 norm, whereas an L^∞ norm can be used to impose the constraint of an error corridor.

An inverse problem of this kind may be represented diagrammatically as in Fig. 2. The two stages in progressing from $D(z)$ to $x(\theta)$ are illustrated. From D, which will not in general belong to the image $\mathcal{M}(\mathcal{C})$ of $\mathcal{C}$ in S, we look for a nearest point $X^\gamma_{0(?)}$ in $\mathcal{M}(\mathcal{C})$ if such a point exists, and then from $X^\gamma_{0(?)}$ we perform the inverse map to $x_{0(?)} \equiv \mathcal{M}^{-1}(X^\gamma_{0(?)})$. The point $x_{0(?)}$ thus obtained is called a non-regularised quasi-solution. Both of these steps are fraught with potential hazards[2] and must be negotiated with care.

A primary concern must be about the continuity of the map from D to x_0. In determining the nearest point X_0^γ we need to consider the existence and uniqueness of this point and its continuity relative to small changes in D. Then in proceeding from X_0^γ to x_0 we have to examine the continuity of the inverse map $\mathcal{M}^{-1}$. The fact that $\mathcal{M}$ is continuous does not imply continuity of $\mathcal{M}^{-1}$ and in general we may expect that small changes in X_0^γ will lead to large variations in x_0. We shall assume that $\mathcal{M}$ is bijective; otherwise $\mathcal{M}$ might associate more than one point in $\mathcal{C}$ with any given point in S, in which case $\mathcal{M}^{-1}$ would not be a map.

The standard method (Tychonov /1963/) for approaching unstable inverse problems of this kind is to restrict the space $\mathcal{C}$ (sometimes referred to as the *Control Space*) to some compact set $\mathcal{C}_\kappa \subset \mathcal{C}$. Compactness of the set $\mathcal{C}_\kappa$ means that the inverse map $\mathcal{M}^{-1}$ from a point X in $\mathcal{M}(\mathcal{C}_\kappa)$ to $x \epsilon \mathcal{C}_\kappa$ will be continuous too (see, for instance, footnote 19 from Section 4). So in-

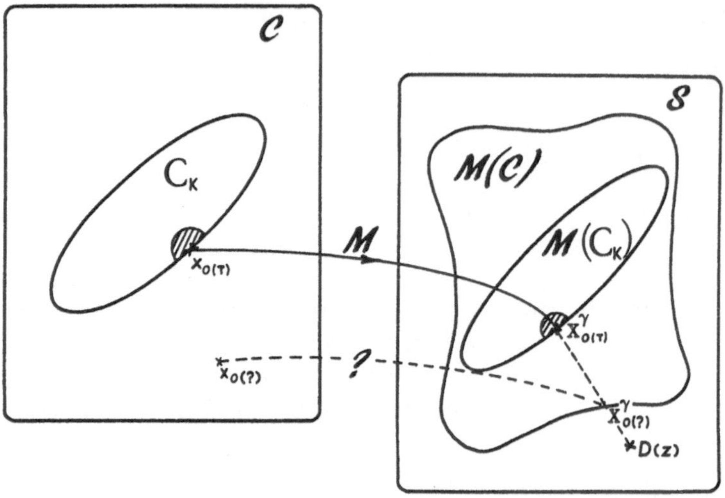

Fig. 2: *The map $\mathcal{M}$, which in the present context is defined by the Poisson integral of eq. (1), is a continuous map from the space C, whose elements are the boundary functions $x(\theta)$, into the space*[1] *S which consists of possible data functions $D(z)$ (real continuous functions on γ) and includes the harmonic functions (restricted to γ)$X^{\gamma}(z)$. $\mathcal{M}(C) \subset S$ is the sub-space of harmonic functions. $X^{\gamma}_{0(?)}$ is the nearest point (assuming that this exists) to D in $\mathcal{M}(C)$ while $X^{\gamma}_{0(T)}$ is the corresponding one in $\mathcal{M}(C_K)$, where C_K is a compact subset of C. Neither $X^{\gamma}_{0(?)}$ nor $x_{0(?)}$ exist in our Poisson case. However the Tychonov regularised quasisolution $x_{0(T)}$ is perfectly well defined.*

stead of proceeding via the nearest point $X^{\gamma}_{0(?)} \epsilon \mathcal{M}(C)$, which may not exist, one looks for the point $X^{\gamma}_{0(T)}$ in $\mathcal{M}(C_\kappa)$ which is nearest to D. In fact the compactness of C_κ implies that $\mathcal{M}(C_\kappa)$ is also compact, which ensures the existence of a nearest point $X^{\gamma}_{0(T)}$. The uniqueness of this point will depend on the geometry of $\mathcal{M}(C_\kappa)$: in particular, convexity of $\mathcal{M}(C_\kappa)$ will guarantee uniqueness. Provided that we have at least local uniqueness then the compactness of $\mathcal{M}(C_\kappa)$ is sufficient to prove that $X^{\gamma}_{0(T)}$ varies continuously with D. Thus, provided that the geometry of $\mathcal{M}(C_\kappa)$ is such that the nearest point $X^{\gamma}_{0(T)}$ is unique, the use of the compact set $C_\kappa \subset C$ has yielded a uniquely defined continuous map $\mathcal{M}_Q^{-1}$ from S to C. Whereas $\mathcal{M}^{-1}$ only makes sense on the image $\mathcal{M}(C)$ of C in S, $\mathcal{M}_Q^{-1}$, yielding what is known as the Tychonov regularized quasisolution $x_{0(T)}$, is defined over the whole space S:

$$x_{0(T)} = \mathcal{M}_Q^{-1}(D) \equiv \mathcal{M}^{-1} \circ \mathcal{P}_{\mathcal{M}(C_\kappa)}(D) \tag{4}$$

where the "projection" $X_{0(T)} = \mathcal{P}_{\mathcal{M}(C_\kappa)}(D)$ corresponds, as shown, of finding the nearest point $X_{0(T)}$ from the set $\mathcal{M}(C_\kappa)$ to D; since $X_{0(T)} \epsilon \mathcal{M}(C_\kappa) \subset \mathcal{M}(C)$, the operation $\mathcal{M}^{-1}(X_{0(T)})$ makes sense. The map $\mathcal{M}_Q^{-1}$ is of course, dependent on the choice of compact set C_κ and one has to ask whether this restriction of C to C_κ is quite arbitrary or can be justified as a constraint which is defined by specific physically determined conditions.

Unfortunately, the physical conditions which are relevant and which one might wish to incorporate as constraints can not typically be expressed in terms of a restriction of the control space C to a compact set C_κ. The sort of conditions which do arise in usual cases,[3] such as bounds, may more readily be cast, by a suitable choice of metric, in the form of a norm constraint $\|x\| \leq \delta$ (or $\|x - \overline{x}\| \leq \delta$). So instead of a restriction of C to a compact set C_κ, the actual restriction based on physical criteria is to a closed ball B_δ, where δ is some finite radius. Unless the control space C is finite-dimensional, in which case the problem of non-compactness does not arise, such a delta-ball is not a compact set and the arguments for the continuity of the inverse map defined by the construction outlined above do not hold. Nevertheless, ball-restrictions of the control space are frequently used as a means to stabilize inverse mappings. The main purpose of this paper is to study the effect of such restrictions and to determine under what circumstances, if any, these can lead to continuous inverse maps.

2. Weak-star compactness

2.1 The Arzelà-Ascoli theorem and relative compactness of $\mathcal{M}(B_\delta)$

The discussion in this Section is based on two important theorems. The first, the Arzelà-Ascoli theorem, states that any infinite family of uniformly bounded equicontinuous functions $\psi_n(s)$, defined on a finite closed interval γ, will have a subsequence $\{\psi_{n'}\}$ which converges uniformly on γ:

$$\psi_{n'}(s) \longrightarrow \tilde{\psi}_0(s). \tag{5}$$

The function $\tilde{\psi}_0$ need not belong to the family ψ_n; thus the Arzelà-Ascoli theorem ensures only relative compactness. To see the relevance of this theorem, consider a linear map $\mathcal{M}$ $\quad(\mathcal{M}(\phi) = \psi)$ of the form

$$\psi_n(s) = \int_\Gamma K(s,t)\phi_n(t)dt \tag{6}$$

where $K(s,t)$ is continuous in s in the closed interval γ, and, for any value of s, $|K(s,t)| < M$ for some finite M. Suppose that the functions $\phi_n(t)$ $\quad(\phi_n \epsilon C)$ are $L^p(1 \leq p \leq \infty)$ on Γ and that they lie within a ball of some finite radius $\delta, B_\delta \subset C$, i.e. $\|\phi_n\|_{L^p} \leq \delta$. Then since $K(s,t)$, being bounded, is L^q in t for any fixed value of s, where $p^{-1} + q^{-1} = 1$, Hölders inequality shows (a) that the moduli of the functions $\psi_n(s)$ are equi-bounded (i.e. they have a bound which is independent of n) and (b), because of the continuity in s of $K(s,t)$, that the $\psi_n(s)$ are also equi-continuous functions.[4] So the conditions of the theorem are satisfied and there is therefore a subsequence $\{\psi_{n'}\}$ of the

$\{\psi_n\}$ which converges uniformly as in eq. (5). However the limit function need not in general be the transform of any function $\phi_0(t)$ and so need not be in $\mathcal{M}(B_\delta)$. One proves in this way[5] that the closure $\overline{\mathcal{M}(B_\delta)}$ is compact although the set $\mathcal{M}(B_\delta)$ itself might not be so: such a set is said to be relatively compact.

The Poisson kernel $P(z,\theta)$ (eq. (2)) is bounded and is continuous in z for $|z| < 1$ and so it satisfies the conditions of the above theorem. Relative compactness of $\mathcal{M}(B_\delta)$ is however not sufficient to yield a nearest point. There will be a nearest point within $\overline{\mathcal{M}}(B_\delta)$ but this may well be on the boundary and thus not belong to $\mathcal{M}(B_\delta)$, in which case the inverse point will not lie in B_δ and indeed may not exist at all.

We have thus to ask under what circumstances, if at all, the set $\mathcal{M}(B_\delta)$ will be compact and not just relatively so. Compactness of $\mathcal{M}(B_\delta)$ will ensure the existence, within $\mathcal{M}(B_\delta)$, of a nearest point; it will also, when combined with the convexity of $\mathcal{M}(B_\delta)$ (which is guaranteed for any linear map such as that of eq. (6) since the ball B_δ is convex [6]), ensure the uniqueness and continuity of that nearest point. We will focus our attention on the possibility of compactness of $\mathcal{M}(B_\delta)$, but in doing so it is important to remember that even if $\mathcal{M}(B_\delta)$ is compact, since B_δ itself is not compact the continuity of the inverse map $\mathcal{M}^{-1}$ from $\mathcal{M}(B_\delta)$ to B_δ is not guaranteed; this is a problem to which we shall return later.

2.2 The Alaoglu-Banach theorem

To make progress we now turn to a second theorem due to Alaoglu and Banach. Before stating this theorem it is necessary to introduce the concepts of weak convergence and weak-* convergence. A sequence $\{u_n\}$ of points in a Banach space U is said to converge weakly to a limit point u if for all linear functionals $< l,. >$ acting on U, i.e. for all elements of the dual space U', $< l, u_n >$ converges to $< l, u >$ for some point $u \epsilon U$. We write this as

$$u_n \rightharpoonup u \quad \Leftrightarrow \quad < l, u_n > \quad \longrightarrow \quad < l, u >, \quad \forall < l,. > \epsilon U'. \tag{7}$$

For weak-* convergence to be defined on a Banach space U it is necessary that U should itself be the dual of another Banach space V. Thus $U = V'$ and we have the inclusion property that $V \subset V'' \equiv U'$. Weak-* convergence is then defined as follows

$$u_n \rightharpoonup^* u \quad \Leftrightarrow \quad < l, u_n > \quad \rightarrow \quad < l, u > \quad \forall < l,. > \epsilon V. \tag{8}$$

The important point to note is that the convergence of functionals need only hold for those functionals l which belong to V, which is a subspace of the dual space U'. Consequently, weak-* convergence is a still weaker condition than weak convergence,[7] unless the space V is reflexive[8] in which case weak and weak-* convergence are the same.

We are now in a position to introduce the Alaoglu-Banach theorem. This theorem states that if U is the dual V' of some Banach space V, then the unit ball B_1 in U (or any ball B_δ of finite radius δ) is compact in the weak-* topology. Compactness of a set $B \subset U$ in the weak-* topology simply means that any sequence $\{u_n\}$ in B has a sub-sequence $\{u'_n\}$ which is weak-* convergent to a point $u \epsilon B$. If the space U is reflexive then $V = U' = V''$, and in this case the Alaoglu-Banach theorem says that the unit ball will be weakly compact. As has already been emphasised (see footnote(7)) weak, or weak-*, convergence does not imply convergence in the norm topology: the ball $B_\delta \subset \mathcal{C}$ is not compact in the norm topology, unless $\mathcal{C}$ is finite dimensional. So, on the one hand, the theorem states that for any sequence $\{u_n\}$ in the ball $B_\delta \subset \mathcal{C}$ a limit point u does exist. However, this is a limit point only in the weak-* (or weak) topology and not in the norm topology which means that in the norm topology u may be far distant from any points of the sequence.

2.3 Compactness of $\mathcal{M}(B_\delta)$

To see what the Alaoglu-Banach theorem has to say about the compactness of the image set $\mathcal{M}(B_\delta) \subset S$ we return to the linear map defined in eq. (6). The functions $\phi_n(t)(\phi \epsilon \mathcal{C})$ are assumed to be $L^p (1 \leq p \leq \infty)$. Now any L^p space, with the single exception of L^1, is the dual of another Banach space L^q, where $p^{-1} + q^{-1} = 1$. L^1 is not the dual of any Banach space. So provided that we exclude L^1 we can say that for any other $L^p (1 < p \leq \infty)$ the Alaoglu-Banach theorem will apply and the ball B_δ will be weak-* compact. So the infinite sequence of functions $\{\phi_n(t)\}$ will have a subsequence $\{\phi_{n'}(t)\}$ which is weak-* convergent to a function $\phi_0(t)$. In particular, since the kernel is bounded, and thus for any fixed value of s $K(s,t)$ as a function of t is L^q, it can be viewed as a linear functional $\langle l_S, \cdot \rangle \epsilon V$ acting on the ϕ_n's, and so, for particular value of $s \epsilon \gamma$, the subsequence of numbers

$$\psi_{n'}(s) = \int_\Gamma K(s,t)\phi_{n'}(t)dt \tag{9}$$

will converge, according to eq. (8), and to the Alaoglu-Banach theorem, to the number

$$\psi_0(s) = \int_\Gamma K(s,t)\phi_0(t)dt, \phi_0 \epsilon B_\delta. \tag{10}$$

The functions $\psi_{n'}(s)$ therefore converge pointwise (for each $s \epsilon \gamma$) to a limit function $\psi_0(s)$ which *belongs* to $\mathcal{M}(B_\delta)$.

Non-uniform pointwise convergence of $\psi_{n'}$ to $\psi_0(s)$ is not itself a proof of compactness[9]: for $\mathcal{M}(B_\delta)$ to be compact we need to show that the $\psi_{n'}$ will converge also in norm to ψ_0. To prove this norm convergence we combine the results of our two theorems. According to the Arzelà-Ascoli theorem the above defined subsequence $\{\psi_{n'}\}$ has a sub-subsequence $\{\psi_{n''(s)}\}$ which converges uniformly on γ to some function $\tilde{\psi}_0(s)$, which although it belongs to the closure $\overline{\mathcal{M}(B_\delta)}$, need not be an element of $\mathcal{M}(B_\delta)$ itself.

However, since uniform convergence implies pointwise convergence, for each fixed value of $s \epsilon \gamma$ the numbers $\psi_{n''}(s)$ will converge both to the number

$\tilde{\psi}_0(s)$ and $\psi_0(s)$. This means that[10] $\tilde{\psi}_0(s) \equiv \psi_0(s)\epsilon\mathcal{M}(B_\delta)$ and so the set $\mathcal{M}(B_\delta)$ is compact in the sense of the L^∞ norm. On the other hand since uniform (L^∞) convergence implies convergence in any other L^p norm, $1 \leq p \leq \infty$, the image $\mathcal{M}(B_\delta)$ of the δ-ball will be compact in all the L^p norms.

The compactness of $\mathcal{M}(B_\delta)$ means that any function, say D, in S has a uniquely defined[11] nearest point ψ_0 in the set $\mathcal{M}(B_\delta)$, which depends continuously on D. Corresponding to ψ_0 there will be a well-defined function $\phi_0\epsilon B_\delta$. However, because B_δ is not compact (in the sense of the norm; it is, as we have seen, only weak-* compact) ϕ_0 will not depend continuously on ψ_0 in the sense that points ϕ in B_δ which are far distant from ϕ_0 will map into points ψ in $\mathcal{M}(B_\delta)$ which are arbitrarily close to ψ_0.[12] This problem will be treated in some detail in Section 3.

2.4 The L^2, L^2 case

If the space $\mathcal{C}$ is L^2 and if our interest in S is for the set $\mathcal{M}(B_\delta)$ to be compact only in the sense of the L^2 norm, then the conditions imposed on $K(s,t)$ may be relaxed. It is sufficient in this case that $K(s,t)$ should satisfy the Hilbert-Schmidt condition:

$$\int_\gamma ds \int_\Gamma dt |K(s,t)|^2 < M \tag{11}$$

where M is some finite bound. This is a much weaker constraint than the requirement that $|K(s,t)|$ be bounded; also, the previous condition that $K(s,t)$ be continuous in s is no longer necessary. Any kernel satisfying the Hilbert-Schmidt condition defines a *compact* operator in the L^2 topology, that is by definition an operator which maps δ-balls into relatively compact sets.[13] Further $\mathcal{C}$ being L^2, is a reflexive space ($\mathcal{C}'' = \mathcal{C}$, as is any L^p except for $p = 1$ or $p = \infty$), weak compactness and weak-* compactness are equivalent so that any sequence $\{\phi_n\}$ in B_δ will have a sub-sequence $\{\phi_{n'}\}$ such that $< l, \phi_{n'} > \rightarrow < l, \phi_0 >$ for any linear functional $< l, . > \epsilon\mathcal{C}'$ acting on the ϕ's.

Consider a linear functional $<< L, . >> \epsilon S'$, the adjoint space of S, which acts on the ψ's; then

$$\begin{aligned} << L, \psi_{n'} >> - << L, \psi_0 >> &= << L, (\mathcal{M}(\phi_{n'}) - \mathcal{M}(\phi_0)) >> \\ &= < \mathcal{M}'L, (\phi_{n'} - \phi_0) > \end{aligned} \tag{12}$$

where $\mathcal{M}'$ is the Banach-space-adjoint-operator of $\mathcal{M}$. Now $< \mathcal{M}'L, . >$ is a linear functional acting on the ϕ's which means that $< l', . > \equiv < \mathcal{M}'L, . > \epsilon\mathcal{C}'$. Also, since $\{\phi_{n'}\}$ converges weakly in $\mathcal{C}$ to ϕ_0 and since eq. (12) holds for any $<< L, . >>$ in S', it follows that the sequence $\{\psi_{n'}\}$ converges weakly in S to the limit $\psi_0\epsilon\mathcal{M}(B_\delta)$. However $\{\psi_{n'}\}$ must also converge to ψ_0 in norm, for if not there will be some ϵ and a sub-sequence $\{\psi_{n''}\}$ of $\{\psi_{n'}\}$ for which $\|\psi_{n''} - \psi_0\| \geq \epsilon$. Since the corresponding subsequence $\{\phi_{n''}\}$ lies within B_δ and $\mathcal{M}$ is a compact operator, $\{\psi_{n''}\}$ must have a further subsequence $\{\psi_{n'''}\}$ which converges in norm to a function $\tilde{\psi}_0 \neq \psi_0$. ($\tilde{\psi}_0$ cannot equal ψ_0 since

$\|\psi_{n''} - \psi_0\| \geq \epsilon$; note that $\tilde{\psi}_0$ need not lie in $\mathcal{M}(B_\delta)$ since in general the range of the compact operator $\mathcal{M}$ is only relatively compact.) The sequence $\{\psi_{n'''}\}$ must also converge weakly to $\tilde{\psi}_0$, but this is impossible since $\{\psi_{n'}\}$ converges weakly to $\psi_0 \neq \tilde{\psi}_0$. Hence $\{\psi_{n'}\}$ must converge in norm (the L^2 norm) to ψ_0, and thus we have shown that $\mathcal{M}(B_\delta)$ is compact in the sense of the L^2 norm when $\mathcal{C}$ is L^2 and the kernel $K(s,t)$ is Hilbert-Schmidt (eq. 11).

Note that this proof of compactness holds with greater generality for any reflexive space $\mathcal{C}$ and compact map $\mathcal{M}$. So the above discussion applies equally well to this more general case.

3. Instability of the inverse map from $\mathcal{M}(B_\delta)$ to B_δ

3.1 Resumé of previous results

The main results of the previous section may be summarized in terms of two alternative sufficient conditions, **A** and **B** below, covering two partially different conditions:

(**A**) the space $\mathcal{C}$ is the dual of a space V to which the kernel $K(s,t)$, considered as a function of t, belongs; also $K(s,t)$ is continuous in s in the closed interval γ.[14]

(**B**) the space $\mathcal{C}$ is reflexive and the map $\mathcal{M}$ is a compact operator. [15] (In both cases we suppose that the map $\mathcal{M}$ is a bijection between $B \subset \mathcal{C}$ and $\mathcal{M}(B) \subset \mathcal{S}$).

If either of these conditions is satisfied and if B_δ is any finite ball in $\mathcal{C}$ then:
(i) $\mathcal{M}(B_\delta)$ is a compact set: condition **A** ensures the compactness in the L^∞ and so in any other L^p-norm for S, while the condition **B** makes predictions for those topologies of $\mathcal{S}$ for which $\mathcal{M}$ represents a compact operator. This means among other things that to any function D in S, there corresponds a nearest point ψ_0 within $\mathcal{M}(B_\delta)$ which depends continuously on D.
(ii) corresponding to ψ_0 there will be a well-defined function $\phi_0 \epsilon B_\delta \subset \mathcal{C}$. A subscript B may be added ($\phi_{0(B)}$) to distinguish it from the Tychonov quasisolution $\phi_{0(T)}$ which is based, as shown, on compact subsets $\mathcal{C}_k \subset \mathcal{C}$ rather than balls.

In contrast with what happens in the Tychonov case, although the function $\phi_{0(B)}$ is well-defined, the map $\mathcal{M}^{-1}$ from $\mathcal{M}(B_\delta)$ to B_δ is not continuous in the norm topology of $\mathcal{C}$, which means that points in $\mathcal{M}(B_\delta)$ arbitrarily close to ψ_0 will map to points of B_δ which are distant from $\phi_{0(B)}$.

In what follows we shall assume that either one of the two conditions **A** or **B** is satisfied. The Alaoglu-Banach theorem tells us that B_δ is weak-* compact (weak-compact in case **B**) which means that any infinite sequence $\{\phi_n\}$ in B_δ has a sub-sequence $\{\phi_{n'}\}$ which is weak- * convergent to a limit $\phi_0 \epsilon B_\delta$. As we have seen in the preceding section the image functions $\psi_{n'} = \mathcal{M}(\phi_{n'})$ from the space S will converge in norm to $\psi_0 = \mathcal{M}(\phi_0)$ but the initial sequence $\{\phi_{n'}\}$, although weak-* convergent to the limit $\phi_0 \epsilon B_\delta$ will not, in general, be convergent in the norm topology of $\mathcal{C}$. This means that for any N, however large, there may be values of $n' > N$, say n'_1, n'_2, for which $\phi_{n'_1}$ and $\phi_{n'_2}$ will be remote from each other and from ϕ_0.

To gain more insight into the mechanism of the possible convergence or nonconvergence of these sequences from B_δ, we return to the analytic continuation problem from §1. For definiteness we shall suppose that the space $\mathcal{C}$ is L^2 and so define the ball B_δ by

$$\int_0^{2\pi} |x(\theta)|^2 d\theta \leq \delta. \tag{13}$$

Further, we shall consider S to be either the space L^∞ or the space L^2 of functions defined on the segment γ where the data function $D(z)$ is also given. Since the space $\mathcal{C}$ of the boundary functions $x(\theta)$ is now L^2 and hence reflexive, and since the equations (1) and (2) define a compact operator from $\mathcal{C}$ to S (the image $\mathcal{M}(B_\delta)$ in S of the ball B_δ through $K(z,\theta) = P(z,\theta)$[16] is relatively compact in either theL^2 or L^∞ choice for the norm of S), the condition **B** is fulfilled and so, according to (i), the set $\mathcal{M}(B_\delta)$ will also be compact. Of course, if $\mathcal{M}(B_\delta)$ is compact in the L^∞ norm, it is automatically so in any other L^p norm.

As has been shown, the compactness of $\mathcal{M}(B_\delta)$ ensures the existence of a best real holomorphic function $X_0(z)$, whose restriction $X_0^\gamma(z)$ on γ is the best approximant of the data $D(z)$ among all the real analytic functions whose boundary values $x(\theta) \equiv ReX(e^{i\theta})$ are subjected to the bound (13). This is true whether the best approximation on γ is defined in terms of the L^2 or the L^∞ norm - but, of course, the results are different, since the two norms define two separate problems. The corresponding boundary function $x_0(\theta) = ReX_0(e^{i\theta})$ represents the quasi-solution $x_0 \equiv \phi_{0(B)}$ discussed in point (ii) above. It is perfectly well defined, and equations can be derived[17] to determine it directly from the data $D(z)$ given on γ. The function $x_0(\theta)$ may then be used to compute $X_0(z)$ at other points in the unit disk.

3.2 To a small subset around X_0^γ corresponds a large *nebula* in $\mathcal{C}$

The point which we would like to discuss here is the existence in the neighbourhood of X_0^γ of some functions $\tilde{X}_n^\gamma \epsilon \mathcal{M}(B_\delta)$, which, being very close to X_0^γ are almost as good approximants of the data D on γ, but whose boundary values $\tilde{x}_n(\theta) \equiv Re\tilde{X}_n(exp(i\theta))$ differ considerably among themselves, being spread over large regions of the ball B_δ.

To this end let us first suppose that $\{x_n\}$ is a sequence from B_δ converging in norm towards x_0. The corresponding X_n^γ converge towards X_0^γ. We shall now prove the following:

(i) For any arbitrarily small positive η one can find a sequence of functions $\tilde{X}_n^\gamma(z)\epsilon\mathcal{M}(B_\delta)$ so that $||X_n^\gamma - \tilde{X}_n^\gamma|| < \eta$, converging to the same limit X_0^γ as does the original sequence $\{X_n^\gamma(z)\}$, but whose corresponding sequences

$\{\tilde{x}_n(\theta)\}$ in B_δ do not converge in norm. (They will converge only in the weak-star sense.)

(ii) Conversely take any norm non-convergent sequence $\{\tilde{x}_n(\theta)\}$ in $B_\delta \subset \mathcal{C}$ which converges in the weak-star sense towards x_0 so that the corresponding $\tilde{X}_n^\gamma$ in $\mathcal{M}(B) \subset S$ converge to $X_0^\gamma(z)$. One can find B_δ-sequences $\{x_n(\theta)\}$ converging strongly to the limit $x_0(\theta)$ such that the corresponding sequences $\{X_n^\gamma(z)\}$ in $\mathcal{M}(B_\delta)$ satisfy the condition $||\tilde{X}_n^\gamma - X_n^\gamma|| < \eta$. In other words, in each neighbourhood[18] of the $X_n^\gamma(z)$ one can find a function $\tilde{X}_n^\gamma$ and viceversa.

The proof of (i) follows from the fact that γ contains only internal points of the unit disk. Since $|z^\gamma| < 1$, for all $z^\gamma \epsilon \gamma, z^n$ can be made as small as one would wish on γ, by taking n to be sufficiently large, while its value $e^{in\theta}$ on the boundary remains finite. By choosing a positive real sequence $\{a_n\}$ such that $a_n \to 1$ for large n while remaining small for n less than some sufficiently large N, so that $|z|_\gamma^N < \eta$, we can define two sequences $X_n^\gamma(z)$ and $\tilde{X}_n^\gamma(z) = X_n^\gamma(z) + a_n z^n$, which will be arbitrarily close on γ but differ on the boundary in such a way that if $x_n(\theta) \to x_0(\theta)$, the sequence $\tilde{x}_n(\theta) = x_n(\theta) + a_n \cos n\theta$ will not converge.

To prove (ii) we make use again of a suitably chosen sequence $\{a_n\}$, small for n less than some N and tending to 1 with large n. Then if the sequence $\{\tilde{x}_n(\theta)\}$ is weak-star but *not* norm-convergent towards $x_0(\theta)$ one can define a new sequence $\{x_n(\theta)\}$

$$x_n(\theta) = \tilde{x}_n(\theta) - a_n(\tilde{x}_n(\theta) - x_0(\theta)), \qquad (a_n \to 1), \tag{14}$$

where obviously $x_n(\theta) \to x_0$. By hypothesis $\{\tilde{X}_m^\gamma\}$ converges to X_0^γ, hence given η, there exists a value for N such that for $n > N$, $||\tilde{X}_n^\gamma - X_0^\gamma|| < \eta$. Since from equation (14), $X_n^\gamma(z) - \tilde{X}_n^\gamma(z) = a_n(\tilde{X}_n^\gamma(z) - X_0^\gamma(z))$, taking for $n \leq N \quad a_n = \eta/||\tilde{X}_n^\gamma(z) - X_0^\gamma(z)||$ and for $n > N \quad a_n \equiv 1$ we get, as desired, $||X_n^\gamma(z) - \tilde{X}_n^\gamma(z)|| < \eta$ independently of n.

We have hence proved that the sequences $\{X_n^\gamma(z)\}$ and $\{\tilde{X}_n^\gamma(z)\}$ from $\mathcal{M}(B_\delta)$ corresponding to convergent and not convergent boundary values respectively, lie densely among themselves. These results are illustrated in Figure 3. However, the continuation (eq(3)) from D to x_0 via the nearest point X_0^γ has apparently been implemented with a uniquely determined result x_0, yet this result is highly unstable in so far as an arbitrarily small change $X_0'^\gamma$ of X_0^γ can produce a large change in x_0'. Given this instability of x_0' with respect to $X_0'^\gamma$ there can be no justification in choosing x_0, out of the extended *nebula* of points in B_δ which map into a small neighbourhood of X_0^γ, as the preferred result to the continuation problem.

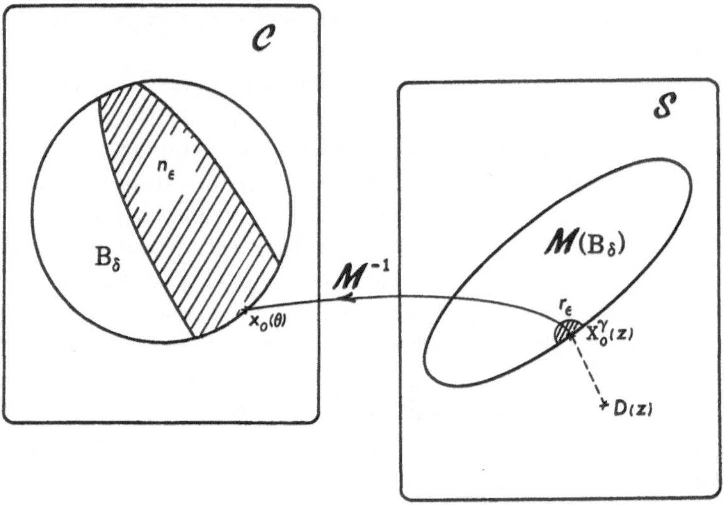

Figure 3. *The small region r_ϵ about X_0^γ maps into an extended nebula in B_δ*

3.3 An alternative definition of stability

We have to conclude that the restriction of $\mathcal{C}$ to a ball B_δ does not have the effect of stabilizing the analytic continuation from γ to the boundary $|z| = 1$. This result is clear; what may be confusing about it is that in spite of this instability, x_0 may nevertheless be stable (continuous) with respect to small changes in D. In section 5 we shall show that for the $L^2 \to L^2$ problem x_0 does depend continuously on D, for a data point D outside the set $\mathcal{M}(B_\delta)$, even though it is unstable with respect to changes in X_0^γ. Despite this possibly surprising result, the instability with respect to the points $X_0'^\gamma \epsilon \mathcal{M}(B_\delta)$ from the neighbourhood of X_0^γ remains and so x_0 cannot be regarded as a stable solution to the analytic continuation problem.

In what follows, a clear distinction will be made between the continuity of the quasisolution $x_0 = \mathcal{M}_Q^{-1}(D)$ with respect to the data D, and its stability: a physically acceptable definition of the latter is (see also eq. (4)):

(a) continuity of $\mathcal{M}_Q^{-1}(D) \equiv \mathcal{M}^{-1} \circ \mathcal{P}_{\mathcal{M}(B_\delta)}(D)$ with respect to $D(\notin \mathcal{M}(B_\delta))$,

(b) continuity of $\mathcal{M}^{-1}$ for the points from $\mathcal{M}(B_\delta)$, so that a small region $r_\epsilon \epsilon \mathcal{M}(B_\delta)$ around $X_0^\gamma = \mathcal{P}_{\mathcal{M}(B)}(D)$ should map into a small region around x_0. As we have seen, the second requirement is not met by the above defined mapping $\mathcal{M}_Q^{-1}$ from S to $\mathcal{C}$.

4. Alternative strategies

4.1. The Spaces $\mathcal{C}$ and S^{out}

In the last section we saw that the imposition of a ball type constraint $\|x\| \leq \delta$ on the space of boundary functions $x(\theta)$ does not lead to a stabilization of the analytic continuation from γ to the boundary $|z| = 1$, even if in some circumstances the map $\mathcal{M}_Q^{-1}(D)$ may be continuous with respect to D (if D is outside $\mathcal{M}(B)$):

$$\mathcal{M}_{Q(B)}^{-1} : D \to X_0^\gamma \to x_0 \text{ (continuous but unstable)} . \tag{15}$$

The weak-* compactness of B_δ (for a ball B_δ defined, let us say, with respect to an L^p norm, $1 < p \leq \infty$, ensures however the following result. For any $\mathcal{L}(w, \theta)$ continuous in θ and bounded (for w in some closed set γ_{out} inside the unit disk) the set of functions defined, for any $x(\theta) \epsilon B_\delta$, by

$$X^{out}(w) = \int_0^{2\pi} \mathcal{L}(w, \theta) x(\theta) d\theta \tag{16}$$

will be compact in the sense of the L^∞-norm. This follows directly from the Arzelà and Alaoglu-Banach theorems, following the same arguments as were used in section 3 which also cover the step from pointwise convergence in w of a sequence $\{X_n^{out}(w)\}$ to norm convergence. Since now both $\mathcal{L}(B_\delta) \subset S^{out}$ and $\mathcal{M}(B_\delta) \subset S^{in} \equiv S$ are compact sets we are almost in the Tychonov case, and so if we can prove that the mapping

$$\mathcal{N} \equiv \mathcal{M} \circ \mathcal{L}^{-1} : \mathcal{L}(B_\delta)(\subset S^{out}) \to \mathcal{M}(B_\delta)(\subset S^{in}) \tag{17}$$

is continuous then[19] the inverse mapping:

$$\mathcal{N}^{-1} \equiv \mathcal{L} \circ \mathcal{M}^{-1} : \mathcal{M}(B_\delta) \subset S^{in} \to \mathcal{L}(B_\delta) \subset S^{out} \tag{18}$$

will be continuous too. (By definition $\mathcal{L}$ and $\mathcal{M}$ and hence $\mathcal{N}$ are bijections: see the conditions **A** and **B** from Section 3.1. In the case of analytic functions, the bijectivity is required by the uniqueness of the analytic continuations).

In fact this problem being "symmetric" with respect to the control space $\mathcal{C}$, the proofs of the continuity of $\mathcal{N}$ or of $\mathcal{N}^{-1}$ are similar in difficulty. We shall hence start directly with the continuity of $\mathcal{N}^{-1}$. From eq. (18) one sees that $\mathcal{N}^{-1}$ is the product of a continuous mapping $\mathcal{L}$ with a discontinuous one, $\mathcal{M}^{-1}$. Hence eq. (18) is not of much help. We can however define a new space $\not{\mathcal{C}}$, whose elements (points) are the same points (boundary functions) $x(\theta)$ from $\mathcal{C}$, but instead of the norm topology it is endowed with the weak-* topology. The point is that the mapping

$$\not{\mathcal{M}}^{-1} : \mathcal{M}(B) \subset S^{in} \to \not{B}_\delta \subset \not{\mathcal{C}} \tag{19}$$

is continuous while at the same time the mapping

$$\not{\mathcal{L}} : \not{B}_\delta \subset \not{\mathcal{C}} \to \mathcal{L}(B_\delta) \subset S^{out} \tag{20}$$

still remains continuous. (Taking a coarser topology in the target space one gets more continuous functions, while when one takes a coarser topology in the source space, there are less continuous functions.) The continuity of $\not\mathcal{M}$ and $\not\mathcal{L}$ is ensured by the arguments of Chapter 2 (by the conditions **A** or **B** and, since the unit ball $\not\mathcal{B}$ is compact in this topology of $\not\mathcal{C}$, the bijectivity of $\not\mathcal{M}$ ($\mathcal{C}$ and $\not\mathcal{C}$ contain the same points) ensures the continuity of $\not\mathcal{M}^{-1}$. Hence

$$\mathcal{N}^{-1} \equiv \mathcal{L} \circ \mathcal{M}^{-1} = \not\mathcal{L} \circ \not\mathcal{M}^{-1} \tag{21}$$

will be continuous. Since moreover $\mathcal{P}_{\mathcal{M}(B)}(D)$ is continuous,

$$\mathcal{N}_Q^{-1}(D) = \mathcal{L} \circ \mathcal{M}^{-1} \circ \mathcal{P}_{\mathcal{M}(B)}(D) : D \to X_0^{\gamma} \to x_0 \to X_0^{out} \tag{22}$$

is continuous too, and so both requirements (a) and (b) from Section 3.3 are satisfied and hence $\mathcal{N}_Q^{-1}$ is stable.

The above conclusions are valid in general for any pair of Banach spaces S^{in} and S^{out} related to $\mathcal{C}$ by operators $\mathcal{M}$ and $\mathcal{L}$ subjected to the conditions **A** or **B**. In the case that S^{in} and S^{out} are the spaces of the restrictions X^{in} and $X^{out}(z)$ of functions analytic in the unit disk to the interior sets $\gamma^{in} \equiv \gamma$ and γ^{out}, one can make the following resumé:

(α) Although the analytic continuation is unique and hence the mapping between S^{out} and S^{in} is a bijection, neither this mapping nor its inverse is continuous.

(β) However, its restriction $\mathcal{N} : \mathcal{L}(B_\delta) \subset S^{out} \to \mathcal{M}(B_\delta) \subset S^{in}$ to the images of the δ ball B_δ from $\mathcal{C}$ (in any $L^p, 1 < p \leq \infty$, topology of $\mathcal{C}$) is continuous, and so is the inverse, $\mathcal{N}^{-1}$.

(γ) The quasisolution $X_0^{out} = \mathcal{N}_Q^{-1}(D) \equiv \mathcal{N}^{-1} \circ \mathcal{P}_{\mathcal{M}(B_\delta)}(D)$, where $X_0^{in} = \mathcal{P}_{\mathcal{M}(B_\delta)}(D)$ is the nearest point from $\mathcal{M}(B_\delta)$ to the data D in S^{in}, is stable, as it satisfies both the requirements (a) and (b) from Section 3.3.

4.2 Examples of control spaces $\mathcal{C}$ and spaces S^{out} in analytic continuation problems

1. As a first example we take $\mathcal{C}$ as before to be the space of the boundary values $x(\theta) \equiv ReX(e^{i\theta})$ on the unit circle, endowed with the L^2 or the L^∞ norm. Let S^{out} be the space of the restrictions $X^{out}(z)$ on the set γ^{out} defined, for instance, to be the disk $|z| \leq \rho$, with $\rho < 1$. The kernel $\mathcal{L}(z,\theta)$ from eq. (16) will then be $P(z,\theta)$, the same as the kernel defining the mapping $\mathcal{M} : \mathcal{C} \to S^{in}$ except that the integration ranges will be different, so the Hilbert-Schmidt condition for $\mathcal{L}(z,\theta)$, is satisfied. The stability of the map $\mathcal{N}_Q^{-1}(D)$, see eq. (22) is in this case the stability of the continuation of $D(z\epsilon\gamma_{in})$, via the nearest point $X_0^{\gamma_{in}}$ from $\mathcal{M}(B_\delta)$ to a function $X^{out}(z)$ holomorphic in the disk $|x| \leq \rho$. As we have seen, this stability is ensured by the boundedness constraint on $ReX(z)$ on the unit circle.

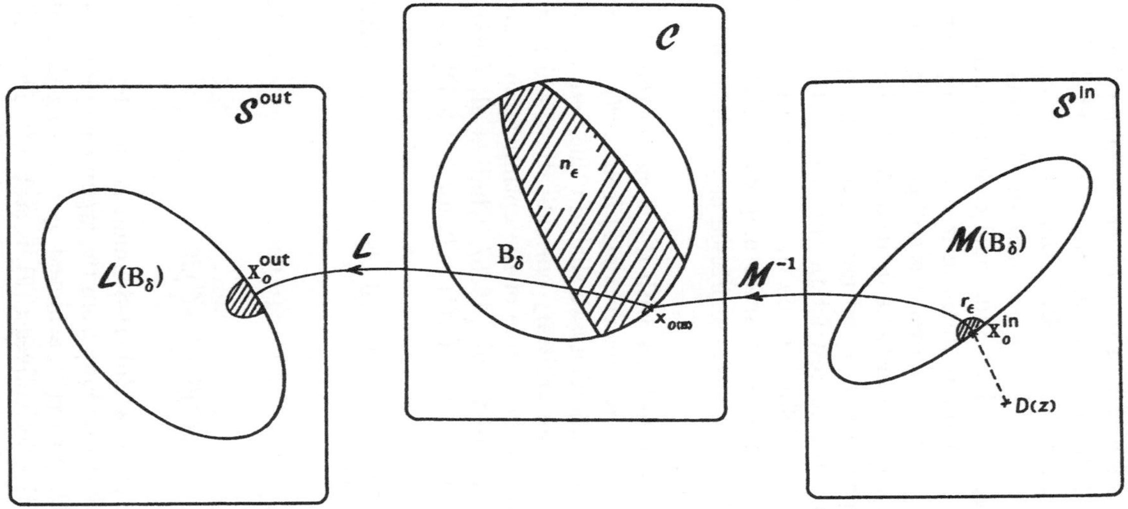

Figure 4. *The mapping* $D \rightarrow X_0^{in} \rightarrow x_{0(B)} \rightarrow X_0^{out}$. *The mapping is stable since the small region* $r_\epsilon \subset \mathcal{M}(B_\delta)$ *around* X_0^{in} *transforms into a corresponding small region in* S^{out}.

2. An obvious comment to make here is that in order to extend this stabilization to $\rho = 1$ one need simply apply the ball constraint $||x|| \leq \delta$ on a circle with radius greater that 1. Unfortunately this option is not normally available; the circle $|z| = 1$ represents the cuts of the s-plane and so is a boundary on which additional constraints such as $||x|| \leq \delta$, may be available and the problem of interest is to perform the continuation to this boundary. However it is still possible to choose an alternative control space C' of functions defined on the boundary $|z| = 1$, which is in a certain sense more remote than the space C. For example if we are prepared to restrict the functions $x(\theta)$ to be differentiable then C' could be defined to be the space of the derivative functions[20] $x'(\theta) \equiv \frac{d}{d\theta} ReX(e^{i\theta})$. A smoothing constraint of the form $\int \sigma(\theta)|x'(\theta)|^2 d\theta, (\sigma(\theta) > 0)$ may be expressed as a ball condition $||x'|| \leq M$.

We shall take C' to be L^2, in accordance with the choice of the above smoothing condition as norm. To re-construct $x(\theta)$ from $x'(\theta)$ we need to normalize at one point. For simplicity we shall do this here by setting $x(0) = 0$; the extension to an arbitrary normalization, which can subsequently be determined in the course of the optimization procedure, is relatively straightforward Ciulli, /1988/, Ciulli et al, /1987/. Then in terms of the usual Θ function ($\Theta(t) = 0$, for $t < 0, \Theta(t) = 1$ for $t \geq 0$).

$$x(\theta) = \int_0^{2\pi} \Theta(\theta - \theta')x'(\theta')d\theta' \tag{23}$$

giving

$$X^\gamma(z) = \int_0^{2\pi} \tilde{P}(z, \theta')x'(\theta')d\theta' \tag{24}$$

where

$$\tilde{P}(z, \theta') = \int_0^{2\pi} P(z, \theta'')\Theta(\theta'' - \theta')d\theta'', \quad z\epsilon\gamma. \tag{25}$$

For $z\epsilon\gamma$ it is clear that $\tilde{P}(z, \theta)$ is a Hilbert-Schmidt kernel. $\Theta(\theta - \theta')$ is also Hilbert-Schmidt, and so it follows from the arguments indicated above that the map from $D(z)$ to $x(\theta)$, achieved by imposing the ball restriction on the space of the derivatives $x'(\theta)$, is stable.[21] This may be seen from Figure 4 by simply re-interpreting C to be C' so that x_0 becomes x_0', while X_0^{out} is now x_0.

5. The L^2 calculation.

5.1. An Integral Equation for the Quasisolution $x_0(\theta)$

Finally, in this section we shall sketch the basis of the actual calculation of $X(z)$ from D, using L^2 norms in both S and C. We shall impose the boundedness condition $||x|| \leq \delta$ on the space of the functions $x(\theta)$ (even though as we have seen this does not define a stabilized continuation to

$|z| = 1$) because we wish to illustrate some features of the map thus obtained. To implement the calculation based on the constraint $\|x'\| \leq \delta$ one has simply to replace P by $\tilde{P}$. For details for these calculations, using in particular the space of derivative functions $\frac{d}{d\theta} Im X(e^{i\theta})$, see Ciulli et al/1984/,/1987/.

The L^2 norm in S is defined in relation to a χ^2 - function

$$\chi^2[X^\gamma] \equiv \|D - X^\gamma\|^2 \tag{26}$$

where

$$\|X^\gamma\|^2 \equiv \int_\gamma (X^\gamma(z))^2 n(z) dz, \tag{27}$$

and the L^2 norm in C is related to the boundedness constraint

$$\|x\|^2 \equiv F_1[x] \equiv \int_0^{2\pi} \sigma(\theta)(x(\theta))^2 d\theta \leq \delta^2. \tag{28}$$

$n(z) \equiv (\epsilon(z))^{-2}$ is an error function associated with the data function $D(z)$, and $\sigma(\theta)$ is an appropriately chosen positive weight function. $X^\gamma(z)$ and $x(\theta)$ are related by equation (1); we use this to write χ^2 as a functional of x which we shall call $F_2[x]$

$$F_2[x] = \int_\gamma \{D(z) - \int_0^{2\pi} P(z,\theta)x(\theta)d\theta\}^2 n(z)dz.$$

We now introduce a Lagrange multiplier λ to define the functional

$$F[x] \equiv F_1[x] + \lambda F_2[x]. \tag{29}$$

The solution $x_0(\theta)$ which we are looking for will yield an extremum of $F[x]$ subject to the subsidiary condition

$$F_1[x_0] \leq \delta^2. \tag{30}$$

Fréchet differentiation of $F[x]$ yields the following Fredholm integral equation of the second kind for $x_0(\theta)$

$$x_0(\theta) = \lambda G(\theta) + \lambda \int_0^{2\pi} K(\theta,\theta')x_0(\theta')d\theta' \tag{31}$$

where the inhomogeneous term $G(\theta)$ and the kernel $K(\theta,\theta')$ are given by

$$G(\theta) = \frac{1}{\sigma(\theta)} \int_\gamma n(z)P(z,\theta)D(z)dz \tag{32}$$

$$K(\theta,\theta') = \frac{-1}{\sigma(\theta)} \int_\gamma n(z)P(z,\theta)P(z,\theta')dz. \tag{33}$$

Note that if we had not introduced the B_δ condition, so that $F[x]$ would simply have been $F_2[x]$, the corresponding result would have been the first kind Fredholm equation

$$G(\theta) + \int_0^{2\pi} K(\theta, \theta') x_0(\theta') d\theta' = 0. \tag{34}$$

So we see that the effect of introducing the B_δ constraint has been to replace the unstable first-kind equation by the stable second-kind Fredholm equation.

At this stage it is convenient to symmetrise the eq. (31). To this end we multiply (31) with $\sigma(\theta)^{\frac{1}{2}}$ and define:

$$\begin{aligned} \tilde{x}_0 &\equiv \sigma(\theta)^{\frac{1}{2}} \cdot x_0(\theta) \\ p(z,\theta) &= P(z,\theta)/\sigma(\theta)^{\frac{1}{2}} \\ g(\theta) &= \int_\gamma n(z) p(z,\theta) D(z) dz \\ k(\theta,\theta') &= -\int_\gamma n(z) p(z,\theta) p(z,\theta') dz \end{aligned} \tag{35}$$

The integral equation (31) now reads[22]

$$\tilde{x}_0(\theta) = \lambda g(\theta) + \lambda \int_0^{2\pi} k(\theta,\theta') \tilde{x}_0(\theta') d\theta' \tag{36}$$

whereas the best approximant $X_0^\gamma(z)$ of $D(z)$ on γ has still a representation of the form of eq. (1):

$$X_0(z) = \int_0^{2\pi} p(z,\theta) \tilde{x}_0(\theta) d\theta. \tag{37}$$

The kernel of eq. (36) is symmetric and Hilbert-Schmidt so let $u_i(\theta)$ and $1/\lambda_i$ be its eigenfunctions and eigenvalues:

$$\int_0^{2\pi} k(\theta,\theta') u_n(\theta') d\theta' = \frac{1}{\lambda_n} u_n(\theta). \tag{38}$$

One of the main goals of this section is to establish the continuity of the 'optimal' solution $x_0(\theta)$ (or $\tilde{x}_0(\theta)$) with respect to the data $D(z)$ (or with respect to $g(\theta)$, which, see eq. (35), is expressed in a stable way with respect to D). To this aim we shall investigate the sign of the Lagrange multiplier λ and of the eigenvalues λ_i. Note that the fact that the integral eq. (36) has the character of a Fredholm equation of the second kind is not sufficient for the establishment of the continuity of $\tilde{x}_0(\theta)$ with respect to $g(\theta)$, since λ, through the δ-ball condition (30), is itself a functional $\lambda = \lambda[g]$ of $g(\theta)$.

5.2. Positivity of λ and negativity of the eigenvalues λ_i

Equation (36) is solved for individual values of λ, giving solutions $x_0 \equiv x_0^\lambda$, and the appropriate value of λ is then determined by the subsidiary condition (30). In the extremum problem the value of λ determines the relative weighting between the two functionals F_1 and F_2: thus as $\lambda \to +\infty$, $F_2[x] \equiv \chi^2$ will tend to zero as the B_δ constraint is relaxed and D is approximated increasingly well by X_0^γ, whereas as $\lambda \to 0$ the χ^2 constraint is relaxed and the function $x_0^2(\theta)$ will tend to zero (see eq. (31)), giving $F_1[x_0] \equiv ||x_0||^2 = 0$. For intermediate positive values of λ, if we define x_0^λ to be the solution to $\delta F[x] = 0$ for the appropriate value of λ in eq. (29) and consider $F_1[x_0^\lambda], F_2[x_0^\lambda]$ as functions $F_1(\lambda), F_2(\lambda)$ of λ, we see from $\delta F[x_0] = 0$, by putting $\delta x_0 = x_0^{\lambda+d\lambda} - x_0^\lambda$, that

$$\frac{dF_1}{d\lambda} + \lambda \frac{dF_2}{d\lambda} = 0. \tag{39}$$

Hence

$$\lambda = -\frac{dF_1}{dF_2}. \tag{40}$$

Increasing the radius δ of B_δ, the set $\mathcal{M}(B_\delta)$ of the functions $X^\gamma(z)$ on γ increases too, and so F_2, which is the minimum of the distance from $D(z)$ to $\mathcal{M}(B_\delta)$, decreases. This means that dF_1/dF_2 is negative and so λ is positive.

To find the sign of the eigenvalues λ_i one observes that the operator $\mathbf{K}$ corresponding to the kernel $k(\theta, \theta')$ is negatively defined. Indeed, if $\tilde{x}(\theta)$ is any L^2 function on the interval $(0, 2\pi)$, one has - cf. eqs. (35) and (37),

$$\int\int k(\theta,\theta')\tilde{x}(\theta)\tilde{x}(\theta')d\theta d\theta' = -\int_\gamma dz n(z) X^2(z) < 0 \tag{41}$$

where

$$X(z) = \int p(z,\theta)\tilde{x}(\theta)d\theta \tag{42}$$

cannot be zero over the whole segment γ (it is a holomorphic function) unless $\tilde{x}(\theta)$ is identically zero. So all eigenvalues λ_i are negative and we have at the same time proved that the ker of $\mathbf{K}$ is void. The result, to which we shall refer later, reads:

$$1 - \lambda_i/\lambda > 1 \quad \text{(strictly greater than 1)} \tag{43}$$

5.3 Continuity of the quasisolution $x_0(\theta)$ with respect to $D(z)$

In order to show that $x_0(\theta)$ is stable with respect to the small variations of $D(z)$, we shall show that the derivatives of $\tilde{x}_i(\theta)$ with respect to the partial waves

$$g_i \equiv \int_0^{2\pi} g(\theta)u_i(\theta)d\theta \tag{44}$$

are all finite. We have hence not only continuity but also differentiability. To this end we project the equation (36) with $u_i(\theta)$ to find

$$\tilde{x}_i = \lambda g_i + \frac{\lambda}{\lambda_i}\tilde{x}_i \tag{45}$$

which yields[23]

$$\tilde{x}_0(\theta) = \sum \tilde{x}_i u_i(\theta) = \sum \frac{\lambda\lambda_i g_i}{\lambda_i - \lambda} u_i(\theta). \tag{46}$$

Taking the derivative of $\tilde{x}_i$ with respect to g_i we have to consider also the dependence of λ on g_i, because of the condition $F_1 = \text{const}$:

$$\delta^2(= \text{ const }) = \int_0^{2\pi} \tilde{x}_0^2(\theta)d\theta = \sum \frac{\lambda_\kappa^2 g_\kappa^2}{(\lambda_\kappa/\lambda - 1)^2}. \tag{47}$$

From

$$\frac{d(\delta^2)}{dg_i} = \frac{\partial(\delta^2)}{\partial g_i} + \frac{\partial(\delta^2)}{\partial \lambda}\frac{d\lambda}{dg_i} = 0 \tag{48}$$

we get

$$\begin{aligned} \frac{d\tilde{x}_i}{dg_j} &= \frac{\partial \tilde{x}_i}{\partial g_j} - \frac{\partial \tilde{x}_i}{\partial \lambda}\frac{\partial(\delta^2)}{\partial g_j} / \frac{\partial(\delta^2)}{\partial \lambda} \qquad (49) \\ &= \frac{\lambda\lambda_i}{\lambda_i - \lambda}\delta_{ij} - \frac{\lambda_i^2 g_i}{(\lambda_i - \lambda)^2} \cdot \frac{2\lambda^2\lambda_j^2 g_j}{(\lambda_j - \lambda)^2} / (2\lambda \sum_k \frac{\lambda_\kappa^3 g_\kappa^2}{(\lambda_\kappa - \lambda)^3}). \end{aligned}$$

From eq. (43) and from $\lambda_\kappa < 0$ we see that all terms from the sum from the r.h. side of eq. (49) are positive and so neither this denominator nor any of the other $(\lambda_j - \lambda)$ denominators can vanish. (The sum in the denominator is convergent since $\sum g_\kappa^2$ is convergent because $g(\theta)$ is $L^2; \sum 1/\lambda_\kappa^2$ is also convergent since the kernel is Hilbert-Schmidt, and so $1/\lambda_\kappa$ has to tend to zero). This means that the r.h.s. of (49) cannot be singular, which proves our assertion.

The explicit expression of $dF_1/d\lambda \equiv \partial(\delta^2)/\partial\lambda$ in terms of λ_i and g_i permits also a direct verification of the Lagrange multiplier identity (39). To this end we compute F_2 as the integral (see also eqs. (37), (33) and (35)):

$$\begin{aligned} F_2 &\equiv \chi^2 = \int_\gamma n(z)\{(D(z) - \int_0^{2\pi} p(z,\theta)\tilde{x}_0(\theta)d\theta)\}^2 dz \\ &= \int_\gamma n(z)D^2(z)dz - 2\int_0^{2\pi} d\theta g(\theta)\tilde{x}_0(\theta) - \int_0^{2\pi}\int_0^{2\pi} k(\theta,\theta')\tilde{x}_0(\theta)\tilde{x}_0(\theta')d\theta d\theta' \\ &= \overline{D^2} - 2\sum g_i\tilde{x}_i - \sum \tilde{x}_i^2/\lambda_i \qquad (50) \end{aligned}$$

where $\tilde{x}_i$ is given in eq. (46). The mean value $\overline{D^2}$ of $D^2(z)$ does not depend on λ, so that one gets finally

$$\frac{dF_2}{d\lambda} \equiv \frac{\delta(\chi^2)}{\delta\lambda} = -2\sum \frac{\lambda_\kappa^3 g_\kappa^2}{(\lambda\kappa - \lambda)^3} \tag{51}$$

and hence the equation (39) is verified identically.

This continuity property of the quasisolution is present (Auberson /1990/, private communication) in all $L^p, 1 < p < \infty$, problems — in the L^∞ case the norm is too 'severe' versus the small changes of $x_0(\theta)$ and so there the solution is discontinuous. As has been emphasized in Section 3, owing to the lack of stability due to the wide spread of the counterimages of small regions around x_0 in C (the "nebula effect") this apparent stability of the point x_0 comes as a surprise and may be indeed very misleading.

5.4. The L^1 problem and the Functions of Bounded Variation

Before concluding we should say something about the L^1 problem. Since the L^1 space is not the dual of any other Banach space, the counter image of the limit function $X_0^\gamma(z)$ might lie outside the unit L^1 ball. Hence the compactness of $\mathcal{M}(B_\delta)$ is not guaranteed and the above analysis does not apply. This is regrettable since L^1 is the most general function space for which the representation (1) is valid. However this function space is imbedded in the more general space of measures based on functions of Bounded Variation, which is the dual of the space $\mathcal{C}$ of continuous functions to which $P(z,\theta)$, for $|z| < 1$, belongs (the Poisson-Schwarz Villat formula is valid also for this measure space). So the condition **A** from Section 3 is satisfied and hence the problem is solvable.

6. Concluding Remarks

It has become common practice in handling inverse problems to use truncated expansions to create compact sets in the control space. Whereas such a use of truncation may be quite acceptable in the computation of results which are stable, it is dangerous when instabilities are present, as will be the case with bounded sets in $\mathcal{C}$ such as the ball B_δ. We have shown in this case that the counterimages x_0' in $\mathcal{C}$ of points X_0' from an n_ϵ small region around the nearest point $X_0 \epsilon \mathcal{M}(B_\delta)$ to the data D, are spread over a large *nebula* in B_δ. The effect of truncation is to select a solution from this extended *nebula* in an essentially arbitrary way.

A source of potential confusion is the fact that in certain cases the so-called optimal function $x_0 \epsilon \mathcal{C}$ corresponding to the nearest point X_0 to the data D, obtained using the boundedness condition $x \epsilon B_\delta$ in $\mathcal{C}$, turns out to be continuous with respect to D. This apparent stability is quite misleading, since any point x_0' corresponding to a point X_0' such that $\|X_0' - D\|$ and $\|X_0 - D\|$ differ by less than ε, must be equally physically acceptable and as we have seen, the points x_0' are spread over an extended region in B_δ. Therefore one must make a distinction between the continuity with respect to the data of the quasisolution $x_{0(B)} = \mathcal{M}_Q^{-1}(D)$ defined with boundedness conditions in $\mathcal{C}$, and its overall stability. As is shown at the end of Section 3, besides the continuity of $\mathcal{M}_Q^{-1}(D)$, stability requires also the continuity of the inverse mapping $\mathcal{M}^{-1}$ from points from $\mathcal{M}(B_\delta) \subset S$ to $\mathcal{C}$, and this latter condition is not met in our case.

Concerning the image set $\mathcal{M}(B_\delta)$ the common wisdom has been that this is a relatively compact set and no more than that. We have shown that if the conditions of the Alaoglu and Arzelà-Ascoli theorems are met - which is the case for most of the integral kernels of interest - the image set $\mathcal{M}(B_\delta)$ is in fact compact. This result is important for the existence of the nearest point X_0 and of the corresponding quasisolution.

Further, the stability which is lacking in the control space $\mathcal{C}$ is restored as shown in Section 4 by separating the output space S^{out} from $\mathcal{C}$. Although the mapping $\mathcal{N}^{-1} = \mathcal{L} \circ \mathcal{M}^{-1}$ from $\mathcal{M}(B_\delta) \subset S^{in}$ to $\mathcal{L}(B_\delta) \subset S^{out}$ is the product of the (continuous) mapping $\mathcal{L}$ with the non-continuous mapping $\mathcal{M}^{-1}$, by altering the topology of the space $\mathcal{C}$, $\mathcal{N}^{-1}$ can be rewritten as a product $\not{\mathcal{L}} \circ \not{\mathcal{M}}^{-1}$ where now both factors are continuous. Hence both requirements from the end of Section 3.3 are met and the stability of the quasi-solution $X^{out} = \mathcal{N}_Q^{-1}(D)$ in S^{out} is established.

Finally we draw attention to the following references: Martin /1964/, Fischer /1981/, Roy /1972/, Valin /1989/, where some of the rigorous bounds which have been derived by André Martin are described.

Acknowledgements

We are grateful to G. Auberson for many helpful discussions, and in particular for first suggesting to us the possible relevance of the Alaoglu-Banach theorem. We have also benefitted from many discussions with G. Mennessier. The on-going contact with P.C. Sabatier and his group in Montpellier, with their wide experience of inverse problems, has been most valuable. One of us, T.D.S., acknowledges support from the CNRS and from the University of Montpellier which enabled him to make several visits to Montpellier during the course of which most of this work was done.

References

Auberson G., Mennessier G. /1989/: Commun Math. Phys.**121**, 49.

Auberson G. /1990/: private communication.

Ciulli M., Ciulli S., Spearman T.D. /1984/: J. Math. Phys. **25**, 3194.

Ciulli S., Geniet F., Mennessier G., Spearman T.D. /1987/: Phys. Rev. **D36**, 3494.

Ciulli M. /1988/: Ph.D. Thesis, Trinity College, Dublin (unpublished).

Conway T.B./1985/: *A Course in Functional Analysis* (Springer, Berlin, Heidelberg, New York).

Fischer J. /1981/: Phys. Rep. **76**, 175.

Martin A. /1964/: *Scattering Theory: Unitarity, Analyticity and Crossing* (Springer, Berlin, Heidelberg, New York).

Roy S.M. /1972/: Phys. Rep. **5C**, 125.

Sabatier P.C. /1987/: *Basic Concepts of Inverse Problems*, in: *Tomography and Inverse Problems* edited by P.C. Sabatier (Adam Hilger, Bristol and Philadelphia).

Tychonov A.N. /1963/: Sov. Math. Dokl. **4**, 1035

Valin P. /1989/: Theoretical Unitary Bounds, Theorems and Scalings for the Strong Interactions at High Energies (McGill University, November 1989) preprint.

Yosida K./1966/: *Functional Analysis* (Springer, Berlin, Heidelberg, New York).

Footnotes

[1]P.C. Sabatier /1987/ uses the notation C and E for these two spaces and this convention has been widely adopted. We have translated the French *Espace* into an English *Space*!

[2]For instance, there is no harmonic function $X^{\gamma}_{O(?)}$ which best approximates a continuous but non-differentiable data function D on γ. The Weierstrass theorem ensures the existence of better and better polynomial approximants $X^{\gamma}_{n}(z)$ (whose boundary values $x_n(\theta)$ differ widely among themselves) but none of these can coincide with D.

[3]Finite dimensional bounded closed sets are compact. But this does not extend to infinite dimensions. For further reading concerning the next two sections, see Conway/1985/ or Yosida /1966/, or the mathematical complements from Appendix E of M. Ciulli /1988/.

[4]Hölders inequality gives:

$$|\psi_n(s)| \leq \Big(\int_\Gamma |K(s,t)|^q dt\Big)^{\frac{1}{q}} \Big(\int_\Gamma |\phi_n(t)|^p dt\Big)^{\frac{1}{p}} \leq M\delta,$$

and

$$|\psi_n(s+\epsilon) - \psi_n(s)| \leq \Big(\int_\Gamma |K(s+\epsilon,t) - K(s,t)|^q dt\Big)^{\frac{1}{q}} \Big(\int_\Gamma |\phi_n(t)|^p dt\Big)^{\frac{1}{p}}.$$

This means that both the functions ψ_n and their continuity modules have bounds which are independent of n, i.e. the functions ψ_n are equibounded and equicontinuous. The notation is somewhat awkward: the kernel $K(s,t)$ is said to be compact although the target space $\mathcal{M}(B_\delta)$ is only relatively compact.

[5]The last point to prove is that sequences of points $\overline{\psi_\kappa}$ from the boundary of $\overline{\mathcal{M}(B_\delta)}$ also have subsequences, with limit points inside $\overline{\mathcal{M}(B_\delta)}$. However since for each point $\overline{\psi_\kappa} \epsilon \overline{\mathcal{M}(B_\delta)}$ one can find a point $\psi_\kappa \epsilon \mathcal{M}(B_\delta)$ with, say, $||\overline{\psi_\kappa} - \psi_\kappa|| < (\frac{1}{2})^\kappa$, and since by the Arzelà-Ascoli theorem the sequence $\{\psi_n\}$ has a subsequence $\{\psi_{n'}\}$ which converges to a limit $\tilde{\psi}_0 \epsilon \overline{\mathcal{M}(B_\delta)}$, $\tilde{\psi}_0$ will also be the limit of the subsequence $\{\overline{\psi}_{n'}\}$ of boundary points.

[6]This will not always, however, be strictly convex, e.g. in L^1 or L^∞.

[7]A sequence $\{u_n\}$ which is weakly convergent to u will not (unless it is also strongly convergent) converge in the norm topology. The concept of weak convergence relates naturally to a weak topology which is defined in terms of neighbourhood bases made up of unions and finite intersections of (cylindrical) sets of the following form

$$N(l_1,\dots l_n;\varepsilon) \equiv \{u : \quad |<l_k,u>| < \varepsilon; k = 1,2,\dots n, n \text{ finite }\}$$

where the linear functionals $< l_k,. >$ range over U'. In terms of the weak topology one can say that u_n tends weakly to u or $u_n \rightharpoonup u$. The weak-* topology is defined in terms of neighbourhood bases of exactly the same form except that in this case the linear functionals $< l_k,. >$ are restricted to V which is smaller than $U' \equiv V''$ (unless U is a reflexive space). If u_n tends to u with respect to the weak-* topology we write this as $u_n \rightharpoonup^* u$.

[8]V is reflexive iff $V = V''$; if V is reflexive then U will also be reflexive.

[9]For example consider the functions $\psi_n(s)$ defined on the interval $[0,1]$ by $\psi_n(s) = ns$ for $0 \leq s \leq \frac{1}{n}$; $\psi_n(s) = 1 - n(s - \frac{1}{n})$ for $\frac{1}{n} \leq s \leq \frac{2}{n}$; $\psi_n(s) = 0$ for $s \geq \frac{2}{n}$. The functions $\psi_n(s)$ are continuous and L^∞; $||\psi_n(s)||_{L^\infty} = 1$ for all n since, by construction $sup\, \psi_n(s) = 1$. For any $s = a > 0$ there is some n_0 $(n_0 > \frac{2}{a})$ so that $\psi_n(s) = 0$ for all $n > n_0$. So here we have an example of a sequence of functions $\psi_n(s)$ converging pointwise to $\psi_0(s) \equiv 0$ for all values of s but not converging in the norm since $||\psi_n(s) - \psi_0(s)||_{L^\infty} = 1$.

[10]Pointwise convergence means that for any fixed value of s and for any $\epsilon > 0$ there exists an N so that, for $n" > N$, we have both $|\psi_{n"}(s) - \psi_0(s)| < \varepsilon/2$ and $|\psi_{n"}(s) - \tilde{\psi}_0(s)| < \varepsilon/2$ i.e. $|\psi_0(s) - \tilde{\psi}_0(s)| < \varepsilon$. Since ε is arbitrary this means that $\psi_0(s) = \tilde{\psi}_0(s)$, and this

being true for all $s\epsilon\gamma$, we have $\tilde{\psi}_0(s) \equiv \psi_0(s)$. Since the functions $\psi_n(s)$ defined by eq. (6) are uniformly bounded (by Hölders inequality theorem, see footnote 4) they are L^∞-space functions.

[11]Distance is, by definition, continuous, and a continuous function always attains its minimum (its extrema) on a compact set. Uniqueness follows from the strict convexity of $\mathcal{M}(B_\delta)$ or of the norm of S - when applicable: remember that the L^∞ and L^1 norms are not strictly convex. The continuity of ψ_0 with respect to D follows from (local) uniqueness and a separate theorem - see the Appendices of M. Ciulli /1988/, Ph.D. thesis.

[12]We shall show nevertheless, in section 5, that the inverse map from D to ϕ_0 defined in this way, may indeed be continuous. However given the discontinuity of ϕ with respect to ψ one has to ask about the physical relevance of this particular choice of ϕ_0. This is discussed further in section 5.

[13]The proof that the linear operator $\mathcal{M}$ defined (eq. (6)) by the Hilbert-Schmidt kernel $K(s,t)$ is compact if the source and target spaces are L^2, proceeds as follows. First transform the two segments γ, Γ to the same interval $[a,b]$. Then eq. (11) says that $K(s,t)$ is L^2 on the product set $[a,b] \times [a,b]$. Let $\{P_i(x)\}$ be an ortho-normal basis on $[a,b]$ and define $K^{(N)}(s,t) \equiv \sum_{i,j=1}^{N} k_{ij} P_i(s) P_j(t)$, where $k_{ij} \equiv \int\int K(s,t) P_i(s) P_j(t) ds dt$. Now the mapping $\mathcal{M}^{(N)}$ defined by the kernel $K^{(N)}(s,t)$ maps B_δ into an N-dimensional space, hence for any N the set $\mathcal{M}^{(N)}(B_\delta)$, being finite-dimensional, is compact. The completeness of the basis $\{P_i\}$ implies that $K^{(N)} \to K$, that is given any $\epsilon > 0$ there exists an N such that for $N' > N, \|K - K^{(N')}\|_{L^2} < \epsilon$. Given a sequence $\{\psi_n\}$ in $\mathcal{M}(B_\delta)$ we may now define an associated sequence $\{\psi_n^N\}$ in $\mathcal{M}^{(N)}(B_\delta)$ such that for any specified ϵ, $\|\psi_n - \psi_n^N\| < \epsilon\delta$. Compactness of $\mathcal{M}^{(N)}(B_\delta)$ tells us that $\{\psi_n^N\}$ has a convergent sub-sequence $\{\psi_{n'}^N\}$ and it follows readily that the sequence $\{\psi_{n'}\}$ is a Cauchy sequence, which, since the function space is complete, converges to a limit ψ_0 (not necessarily within $\mathcal{M}(B_\delta)$). It follows that $\mathcal{M}(B_\delta)$ is a relatively compact set.

[14]Note that since the dual-dual space $V'' = \mathcal{C}$ always contains the initial space $V, K(s,t)$ for any particular s acts as a bounded linear functional on the $\phi's$ so that the boundedness required for the Arzelà-Ascoli theorem is implied by the requirement that $\mathcal{C}$ is the dual V' of a space to which $K(s,t)$, as a function of t, belongs.

[15]i.e., as discussed in Section 2, $\mathcal{M}$ maps bounded sets from $\mathcal{C}$ into relatively compact sets from $\mathcal{S}$. The Arzelà-Ascoli condition - see §2.1 - defines relatively compact sets if $\mathcal{S}$ is L^∞, while if we take both $\mathcal{C}$ and $\mathcal{S}$ to be L^2 then the compactness requirement for $\mathcal{M}$ is that the kernel $K(s,t)$ be Hilbert-Schmidt (as in eq. (11)).

[16]$P(z,\theta)$, for $z\epsilon\gamma, \theta\epsilon\Gamma \equiv [0,2\pi]$, is bounded and is continuous in z, hence the functions in any sequence $\{X_n^\gamma(z)\}$ in $\mathcal{M}(B_\delta)$ are both equi-bounded and equi-continuous. $P(z,\theta)$ is L^2 as a function of θ. On the other hand, since on the segment γ the modulus of z is strictly less than 1, the double integral of $|P(z,\theta)|^2$ over $z\epsilon\gamma$ and $\theta\epsilon\Gamma$ is finite and hence $P(z,\theta)$ is also Hilbert-Schmidt. Thus we see that in either case, whether S is L^∞ or L^2, $P(z,\theta)$ defines a compact operator. Moreover, since in our case $\mathcal{C}$ is reflexive, both the conditions **A** and **B** are satisfied.

[17]For instance, if one choses for $\mathcal{S}$ the L^2-norm topology, this equation for $x_0(\theta)$ has the form of a Fredholm integral equation of the second kind. For further details see Section 5.

[18]Some care is required here. It is important to note that since $\mathcal{M}(B_\delta)$ is compact, it cannot contain any finite ball since this would be non-compact. Hence any ball neighbourhood must extend outside $\mathcal{M}(B_\delta)$ and so when speaking about neighbourhoods we shall understand their intersections with the set $\mathcal{M}(B_\delta)$.

[19]If $\mathcal{N}$ is a bijection between the compact sets A and B and moreover $\mathcal{N}$ is continuous, then $\mathcal{N}^{-1}$ is continuous too; this elementary theorem represents the basis of Tychonov's stabilization procedure with compact sets. Continuous functions are characterised by the fact that they map convergent sequences to convergent sequences. Now let $\{b_n\}$ be (any) convergent sequence from B; B being compact will also contain the accumulation point

b_0. Now since $\mathcal{N}$ is a bijection, $a_n = \mathcal{N}^{-1}(b_n)$ represents a sequence from A, which, since A is compact, has to have at least one accumulation point a'_0. But a'_0 has to coincide with $a_0 \equiv \mathcal{N}^{-1}(b_0)$ (and hence, since this is valid for any convergent sequence $\{b_n\}, \mathcal{N}^{-1}$ has to be continuous) since otherwise, $\mathcal{N}$ being continuous, $b'_0 = \mathcal{N}(a'_0)$ and not b_0 ($b'_0 \neq b_0$ since $\mathcal{N}$ is bijective) would have to be the limit point of $\{b_n\}$, which would contradict our hypothesis.

[20]This example is used here for simplicity of illustration. In problems arising in physics it is usually more appropriate to restrict the oscillation of the imaginary part of the amplitude on the cut. So C' would be chosen as the space of the derivatives $\frac{d}{d\theta}\, ImX(e^{i\theta})$ or equivalently $\frac{\partial}{\partial r} ReX(re^{i\theta})|_{r=1}$. See Ciulli et al, /1984/

[21]The L^∞ case has been studied by Auberson and Mennessier /1989/.

[22]There is a computer code which solves this equation and which provides the necessary computer environment for this kind of problem (Appendix C (Ciulli M, /1988/)) which will be sent on request.

[23]Since the eq. (36) represents an integral representation for the difference $\tilde{x}(\theta) - \lambda g(\theta)$, where the kernel $k(\theta, \theta')$ satisfies the condition **A** from §3.1, and since the function $g(\theta)$ defined in eq. (35) is continuous, the solution $\tilde{x}_0(\theta)$ of the integral equation (36) is not only L^2 but represents at the same time a continuous function from L^∞. The solution of the integral equation is usually written in the form $\tilde{x}_0 = \lambda g(\theta) + \lambda^2 \sum g_i/(\lambda_i - \lambda) \cdot u_i(\theta)$ to also make provision for that part $g_\perp(\theta)$ of $g(\theta)$ which is orthogonal to all eigenfunctions $u_i(\theta)$. Since in our case $ker\mathbf{K}$ is void, $g_\perp(\theta)$ does not exist and one can moreover prove (see for instance Chapter 4 and Appendix B from M. Ciulli, /1988/) that the expansion $\sum \tilde{x}_\kappa u_\kappa(\theta) \equiv \lambda \sum \lambda_i g_i/(\lambda_i - \lambda) u_i(\theta)$ converges, not only in the sense of the L^2 norm but also uniformly, to $\tilde{x}_0(\theta)$

Variational Improvement of Perturbation Theory and/or Perturbative Improvement of Variational Calculations

A. Neveu

Laboratoire de Physique Mathématique, Université Montpellier II, F-34095 Montpellier Cedex 05, France

Abstract

We propose an improvement of standard quantum perturbation theory around the vacuum or some other classical solution, in order to incorporate from the beginning quantum mechanical effects in a non-trivial way into functional methods. This improvement is obtained by introducing an a priori *arbitrary* quadratic kinetic term in the action. At a given order in perturbation, the answer depends on the kernel of this quadratic term. Equations of motion are then obtained by imposing that the answer be stationary with respect to this kernel. In particular lowest order cases, one recovers mean-field theory, variational, Hartree and Hartree-Fock results, for which this method can be seen as providing a systematic improvement. It also provides an approach to bound states, like positronium and the hydrogen atom, which, although present in weak coupling, cannot appear in a standard functional treatment of the classical field theory. Finally, contrary to most previous attempts at variational calculations in continuum quantum field theories, this approach is compatible with renormalization and the usual cancellation of infinities.

1 Introduction

There are several approaches to approximate solutions of the quantum mechanics of systems with a finite number of degrees of freedom. The most fascinating, perhaps, are variational or self-consistent calculations. Their fascination lies in their conceptual simplicity, and often numerically excellent results. They have been used in almost all areas of quantum mechanics. The most notable exception is elementary particles and quantum field theories, if we set aside a few attempts around 1974-1975. Apart from the complication of the infinity of the number of degrees of freedom, this is due to the expected clash between virtually any variational guess and the subtle mechanism of infinity cancellations by counterterms in the most interesting cases of renormalizable theories, as emphasized with pessimism by S. Coleman /1977/: the counterterms are determined by requiring ultraviolet finiteness of perturbation theory order by order, and, indeed the probability of such a cancellation for a more or less randomly chosen trial wave functional seems infinitesimal. Furthermore, there was no natural procedure for a systematic improvement of the trial wave functional. The sad state of the art of variational calculations in quantum field theory is manifest in the proceedings of the Wangerooge workshop (L. Polley and D. Pottinger eds./1988/), particularly in Feynman's contribution.

In this paper, we wish to suggest that this pessimism is not warranted, and that a lot more can be done, and access eventually obtained to phenomena usually considered as non-perturbative in quantum field theory through variatio al calculations that are both compatible with the cancellation of infinities and can be improved perturbatively, i.e. by a procedure very much reminiscent of ordinary perturbation theory. These non-perturbative phenomena include the calculation of condensates and bound states.

The problem of bound states in quantum field theories is particularly fascinating. A standard method has been to use the Bethe-Salpeter equation, which is a systematic improvement of the non-relativistic Schrödinger equation for two-body bound states. Fifteen years ago, a large class of bound states was constructed by quantizing classical solutions with finite energy and finite spatial extension. When they can overlap, like in the case of soliton-antisoliton bound states, semi-classical quantization and the Bethe-Salpeter equation give the same results. However, so far, each of these methods has its own rather restricted application range: the Bethe-Salpeter equation is restricted to two-body bound states, and the semi-classical method requires the possibility of large quantum numbers, thus ruling out fermions, except through the trick of a large number N of fermion species. For example, the hydrogen atom, or positronium, though they are weak coupling bound states, cannot possibly appear as the result of the quantization of some classical solution of the equations of motion of quantum electrodynamics. This impossibility is due to the fact that the Coulomb interaction is repulsive between like charges. It requires for the formation of a bound state a roughly neutral system, and thus rules out large values of the charge, and the existence of a semi-classical limit. In order to reach such bound states by functional methods, an improvement is necessary, which goes beyond the classical equations of motion. If one could only subtract the field corresponding to the self-interaction of the electron and the positron, leaving only their mutual interaction, much would have been achieved. This distinction appears in a formalism which includes one loop effects; some prescription must be given for computing the functional determinant of the quadratic form in the functional integral, which distinguishes filled and empty states.

The basic idea of this paper is very simple: add and subtract an *arbitrary* Gaussian kinetic term in the classical Lagrangian, lump the added piece together with the already present kinetic term, and the subtracted piece together with the interaction terms. Formally, one has done nothing, but a given perturbative order will exhibit dependence over the Gaussian modification. Since to all orders this dependence cancels, one optimizes a given order of perturbation theory by choosing its extremum with respect to the Gaussian.

In several respects, the present work can be seen as a particular application, extension, and improvement of the work of R. Balian and M. Vénéroni/1988/, to a functional Lagrangian, rather than Hamiltonian, framework in order to keep manifest Lorentz invariance and a systematic order by order compatibility with perturbation theory. There is also some overlap with previous attempts at optimizing perturbation theory, in particular the "principle of minimal sensitivity", used for example by A.D. Duncan and M. Moshe/1988/ and H.F. Jones and M. Monoyios/1989/ (and further references therein) but, as should be clear to the reader, our approach departs in several crucial respects from the methods developed in those references.

Finally, we stress the fact that our method has a much wider range of applicability than relativistic quantum field theory, and can be used in all cases where a systematic improvement of mean-field calculations are required. Its physics is quite clear, as already discussed by Balian and Vénéroni/1988/, and emphasized above: we replace the interacting non-linear system by an a priori arbitrary Gaussian free system, and adjust the Gaussian to optimize a given order of perturbation theory. We note, as emphasized by Balian and Vénéroni/1988/, that the resulting equations depend on the quantity which one wants to compute. Although the equations that we obtain for the Gaussian are perturbative, their solution is not. In this paper, we shall see several examples of this phenomenon, which is quite familiar in the calculation and minimization of effective potentials.

2 The case of simple integrals

In this section, we discuss the calculation of simple integrals with our method. It is an amusing exercise by itself, and may provide a feeling for the numerical accuracy of the method. Consider first

$$I = \int_{-\infty}^{+\infty} dx e^{-\beta x^4} = \frac{1}{2}\beta^{-1/4}\Gamma\left(\frac{1}{4}\right) \tag{1}$$

Eq.(1) can be considered as a primitive example of a non-perturbative problem: the integrand has no Gaussian saddle point. Now, it is an elementary exercise to prove the following inequality, valid for all integer n, and real x and Δ:

$$e^{-\Delta x^2}[1 + (\Delta x^2 - \beta x^4) + \ldots + \frac{1}{(2n+1)!}(\Delta x^2 - \beta x^4)^{2n+1}] \leq e^{-\beta x^4} \quad (2)$$

From this inequality, it is natural to consider

$$I_n(\Delta) = \int_{-\infty}^{+\infty} dx e^{-\Delta x^2}[1 + (\Delta x^2 - \beta x^4) + \ldots + \frac{1}{(2n+1)!}(\Delta x^2 - \beta x^4)^{2n+1}] \quad (3)$$

and to try and compute I by replacing it with the maximum value I_n of $I_n(\Delta)$ with respect to Δ, at a given n. Following this procedure, one obtains, for the first few values of n:

$$I_0(\Delta) = \frac{3\sqrt{\pi}}{2\sqrt{\Delta}}\left[1 - \frac{\beta}{2\Delta^2}\right] \quad (4)$$

$$I_1(\Delta) = \frac{7\sqrt{\pi}}{16\sqrt{\Delta}}\left[5 - \frac{27\beta}{2\Delta^2} + \frac{65\beta^2}{4\Delta^4} - \frac{495\beta^3}{8\Delta^6}\right] \quad (5)$$

From (4),(5), we see easily that $I_n\beta^{1/4}$ is a pure number, and this provides a sequence of approximations for $\Gamma(1/4)$. Numerically, one obtains

$$2I_0\beta^{1/4} = 3.3832\ldots \quad (6)$$

$$2I_1\beta^{1/4} = 3.5836\ldots \quad (7)$$

$$2I_2\beta^{1/4} = 3.619\ldots \quad (8)$$

One can prove that this sequence converges to the exact value 3.625... for large n. The accuracy of the first approximation I_0, of about 7 percent, is typical of the usual order of magnitude of Gaussian variational approximations. Inequality (2) can be considered as a systematic improvement of the variational formalism of Balian and Vénéroni.

We could also have considered the integral

$$\int_{-\infty}^{+\infty} dx e^{-\beta x^4 - \frac{1}{2}m^2x^2} \quad (9)$$

which has a Gaussian saddle point, $x = 0$, around which one can do ordinary perturbation theory. Perturbation theory can also be recovered by setting $\Delta = \frac{1}{2}m^2$ in the inequality

$$e^{-\Delta x^2}[1 + (\Delta x^2 - \frac{1}{2}m^2x^2 - \beta x^4) + \ldots + \frac{1}{(2n+1)!}(\Delta x^2 - \frac{1}{2}m^2x^2 - \beta x^4)^{2n+1}] \leq e^{-\beta x^4} \quad (10)$$

However, an answer which can only improve over perturbation theory is obtained by taking the maximum with respect to Δ of the integrated left-hand side of this equation. The rather remarkable result is that while at any fixed Δ the integrated left-hand side ultimately diverges for n large enough, corresponding to the divergence of ordinary perturbation theory, (which is Borel summable only), this divergence disappears when Δ is adjusted to its optimal value $\Delta_0(n)$ as n changes. Indeed, for $n \to \infty$, one can show that

$$\frac{2\beta}{\Delta_0^2} \sim \frac{1}{\alpha n} \tag{11}$$

with

$$\alpha = (\gamma^2 - 1)^{1/2} \tag{12}$$

$$\gamma = \frac{1}{2} ln \frac{\gamma + 1}{\gamma - 1} = 1.19968 \ldots \tag{13}$$

This behavior of the effective dimensionless coupling constant $\frac{\beta}{\Delta_0^3}$ is enough to change the Borel summable divergence of perturbation theory into convergence. The fact that Δ_0^2 increases like n also means that the term Δx^2 ultimately dominates over any finite bare mass $m^2 x^2$ term, if present, in eq.(10), and that the procedure for integral (9) should also converge to the exact answer for non-zero m^2, whether positive *or negative*.

3 Anharmonic oscillator

The usual anharmonic oscillator is the first testing ground for any new non-perturbative idea, and we now follow André Martin in this well-established tradition. There are several types of calculations one can do following the basic idea explained in the introduction of this paper, depending on how much energy and time one is willing to spend on this simple yet fascinating system. The simplest one is to restrict the Gaussian kernel to being a constant. This amounts to a very drastic reduction of the power of the general method, but is nevertheless quite interesting, as we shall see. It also has the enormous advantage of involving almost no calculation beyond the results already present in the literature.

Starting from the Lagrangian

$$L = \frac{1}{2}(\partial_t \phi)^2 - \frac{1}{2} m^2 \phi^2 - \frac{\lambda}{4} \phi^4 \tag{14}$$

we rewrite it as

$$L = L_0 + L_1 \tag{15}$$

$$L_0 = \frac{1}{2}(\partial_t \phi)^2 - \frac{1}{2} \omega^2 \phi^2 \tag{16}$$

$$L_1 = \frac{1}{2}(\omega^2 - m^2)\phi^2 - \frac{\lambda}{4} \phi^4 \tag{17}$$

and do perturbation theory to some finite order in L_1, optimizing the answer with respect to ω. Let us apply this to the ground state energy, whose ordinary perturbative value is

$$E_0 = \frac{1}{2}m + \frac{3}{4}m\frac{\lambda}{4m^3} - \frac{21}{8}m(\frac{\lambda}{4m^3})^2 + \frac{333}{16}m(\frac{\lambda}{4m^3})^3 + \ldots \tag{18}$$

The split of eq.(15) amounts in eq.(18) to writing

$$m = \omega\sqrt{1 + \frac{m^2 - \omega^2}{\omega^2}} \tag{19}$$

and expanding the square root, considering that $m^2 - \omega^2$ is of the same perturbative order as λ. For example, to lowest orders, we obtain the following sequence of approximations

$$E_0^{(1)}(\omega) = \frac{1}{2}\omega + \frac{m^2 - \omega^2}{4\omega} + \frac{3\lambda}{16\omega^2} \tag{20}$$

$$\begin{aligned} E_0^{(3)}(\omega) = \frac{1}{2}\omega &+ \frac{m^2 - \omega^2}{4\omega} - \frac{(m^2 - \omega^2)^2}{16\omega^3} + \frac{(m^2 - \omega^2)^3}{32\omega^5} \\ &+ \frac{3\lambda}{16\omega^2} - \frac{3\lambda(m^2 - \omega^2)}{16\omega^4} + \frac{3\lambda(m^2 - \omega^2)^2}{16\omega^6} \\ &- \frac{21\lambda^2}{128\omega^5} + \frac{105\lambda^2(m^2 - \omega^2)}{256\omega^7} \\ &+ \frac{333\lambda^3}{1024\omega^8} \end{aligned} \tag{21}$$

In (20) and (21), we may go to the extreme case $m = 0$, and, minimizing with respect to ω, we obtain

$$E_0^{(1)}(m = 0) = \lambda^{1/3}0.429\ldots \tag{22}$$

$$E_0^{(3)}(m = 0) = \lambda^{1/3}0.4209835\ldots \tag{23}$$

while the exact value is

$$E_0^{exact}(m = 0) = \lambda^{1/3}0.42080497\ldots \tag{24}$$

We note that $E_0^{(n)}(m = 0)$ automatically exhibits the correct $\lambda^{1/3}$ behavior, and that this typically non-perturbative behavior is obtained through a purely perturbative calculation. This phenomenon already appeared in the previous section. In a future publication, we shall show that it also occurs in renormalizable asymptotically free theories, where it takes the form of dimensional transmutation. We also note the substantial improvement when going from (22) to (23), by roughly a factor 20 in precision, and that $E_0^{(1)}(\omega)$ is *exactly* the same as obtained by approximating the ground state wave function by the Gaussian wave function of the harmonic oscillator with frequency ω. Both $E_0^{(1)}(\omega)$ and $E_0^{(3)}(\omega)$ are always above the exact answer E_0^{exact}, and have only one minimum, displayed in (22) and (23). $E_0^{(2)}(\omega)$ and $E_0^{(4)}(\omega)$ have no real extremum.

The situation becomes puzzling, and all the more interesting, when we go to higher orders, and to higher excited states. Let us first consider $E_0^{(5)}(\omega)$ at $m = 0$. We find that it also has only one minimum $E_0^{(5)}(m = 0)$, whose value is

$$E_0^{(5)}(m = 0) = \lambda^{1/3} 0.4207888\ldots \tag{25}$$

This is agqin a substantial improvement in precision, by a factor of order 20, from $E_0^{(3)}(m = 0)$ given in eq. (23), but, more surprisingly, this value is *below* the exact answer (24), by about 2.10^{-5} in relative value. To order seven, $E_0^{(7)}(\omega)$ at $m = 0$ again has only one minimum, whose value is

$$E_0^{(7)}(m = 0) = \lambda^{1/3} 0.4207872\ldots \tag{26}$$

which is *smaller* (by about $1.6.10^{-6}$ in relative value) than the previous approximate value (25). The sequence of minima thus seems to be converging towards a wrong answer. Why it misses by so little is a puzzle. We note however that $E_0^{(7)}(\omega)$ has two inflection points, with very small values ($\sim 10^{-4}$) of the first derivative with respect to ω, at $\frac{\lambda}{\omega^3} \simeq 0.23$ and $\frac{\lambda}{\omega^3} \simeq 0.24$, while the minimum (26) occurs at $\frac{\lambda}{\omega^3} \simeq 0.178228$. We have not investigated whether the nearby complex extrema whose existence is implied by these real inflection points would yield better values for E_0. Also, we have not investigated higher-order approximations. It might be that the modification used for the kinetic term, introducing a simple harmonic oscillator, and correspondingly a single variational parameter is too crude to give a convergent sequence of approximations. It would be worthwhile to repeat the calculation with more variational parameters involved.

We have also applied the same ideas to higher excited states. For the n-th excited state, the ordinary perturbative answer for the Lagrangian of eq.(14) is, to third order in λ:

$$\begin{aligned} E_n = &\left(n + \frac{1}{2}\right) m + \frac{3\lambda}{16m^2}(2n^2 + 2n + 1) \\ &- \frac{\lambda^2}{256m^5}\left[68\left(n + \frac{1}{2}\right)^3 + 67\left(n + \frac{1}{2}\right)\right] \\ &+ \frac{3\lambda^3}{2048m^8}\left[250\left(n + \frac{1}{2}\right)^4 + 569\left(n + \frac{1}{2}\right)^2 + \frac{513}{8}\right] \end{aligned} \tag{27}$$

Here again, we wish to explore the most extreme case, $n \to \infty, m = 0$. In the limit $n \to \infty$, the WKB approximation provides the exact answer, which, in the $m = 0$ case, reads

$$E_n^{exact} \sim \lambda^{1/3}\left(n+\frac{1}{2}\right)^{4/3} 0.867148\ldots \tag{28}$$

Hence, since the lowest order WKB answer is just a one loop calculation around the classical trajectory, we could use an arbitrary kernel (and linear term, to handle the expected shift of ϕ by the classical trajectory) and discover that to lowest order we do recover the exact WKB answer. Ignoring this more sophisticated possibility, we can instead apply to (27) the same treatment as the one we applied to the ground state, using the identity (19), expanding and extremizing. To lowest order, we find

$$E_n^{(1)} \sim \lambda^{1/3}\left(n+\frac{1}{2}\right)^{4/3} 0.8585357\ldots \tag{29}$$

which is smaller than the exact value by about one percent. We again note that the correct functional dependence on λ *and* n has come out automatically of our perturbative calculation.

In next order, we find, for $n \to \infty$, contrary to the case $n = 0$, that $E_n^{(2)}$ has two extrema, one occuring for $(2n+1)\frac{\lambda}{\omega^3} = 1.511719\ldots$, whose value is

$$E_n^{(2)} \sim \lambda^{1/3}\left(n+\frac{1}{2}\right)^{4/3} 0.867405\ldots \tag{30}$$

and the other occuring for $(2n+1)\frac{\lambda}{\omega^3} = 0.7471044\ldots$, whose value is

$$E_n^{(2)} \sim \lambda^{1/3}\left(n+\frac{1}{2}\right)^{4/3} 0.858236\ldots \tag{31}$$

(31) is slightly worse than (29), but (30) is a spectacular improvement, larger than the exact value (28) by only 3.10^{-4}. In third order, we find similarly that $E_n^{(3)}(\omega)$, for $n \to \infty$, has three extrema, one for $(2n+1)\frac{\lambda}{\omega^3} = 0.47905\ldots$, whose value is

$$E_n^{(3)} \sim \lambda^{1/3}\left(n+\frac{1}{2}\right)^{4/3} 0.8593055\ldots \tag{32}$$

another for $(2n+1)\frac{\lambda}{\omega^3} = 1.164148\ldots$¡ whose value is

$$E_n^{(3)} \sim \lambda^{1/3}\left(n+\frac{1}{2}\right)^{4/3} 0.8677991\ldots \tag{33}$$

and the last one for $(2n+1)\frac{\lambda}{\omega^3} = 1.530136\ldots$, whose value is

$$E_n^{(3)} \sim \lambda^{1/3}\left(n+\frac{1}{2}\right)^{4/3} 0.8670673\ldots \tag{34}$$

which is smaller than the exact answer by only 10^{-5}! This is most remarkable for a calculation based on only third order of perturbation theory, and which contains no free parameter. Note also how close (32), (31) and (29) are, and also (33) and (30). All this calls for further investigations. Here, we only remark that the extremum occuring for the smallest value of ω gives the best approximation. A similar fact has been noticed in a different context by Seznec and Zinn-Justin/1979/.

4 Field Theories

The most interesting field theories are renormalizable. The main problem of a variational approach applied to such theories is to yield finite answers. The Lagrangian involves bare parameters, which are infinite. These infinities are designed to precisely cancel the infinities which occur in ordinary perturbation theory around some standard minimum of the classical action (assuming this perturbation theory makes sense). The infinite counterterms are known to some finite, usually quite small, order in perturbation theory. Any modification of the usual perturbative procedure must be such that it systematically preserves the cancellation of these infinities. Modifying the propagator, as we propose, will achieve this only provided that the modification is soft enough in the ultraviolet region. Indeed, a trivial example would be to precisely use the modification obtained by introducing a box of arbitrary size and shape, with appropriate boundary conditions, adding and subtracting this modification according to the procedure described in the introduction, and optimizing with respect to the size and shape of this artificial box the result of a given perturbative order. All steps of the procedure being finite, the optimal size and shape of the box will automatically come out as finite, order dependent, renormalization group invariant quantities. The problem is only one of numerical convergence of the procedure, as the perturbative order increases.

This example of an artificial box can be considered an existence proof of the possibility of performing non-trivial variational calculations in a renormalizable quantum field theory. It is of course a very crude example, and much more interesting results would certainly be obtained by letting the kernel which modifies the kinetic term be more arbitrary. However, it illustrates an important point: the modifying kernel should not introduce any new divergence, or equivalently, should not require any new counterterm, since any arbitrariness in the finite part of such a counterterm would ruin any predictive power of the method.

It is unlikely that the non-linear integral equations which one obtains when varying with respect to this kernel can be solved in general. One would presumably have to resort to clever restrictions of the functional form of this kernel before optimization. The main condition that these restricted functional forms should satisfy is to preserve the cancellation of infinities; if they don't, it just means that the guess is not soft enough in the ultraviolet, and one has to improve it in that region.

The simplest form of a kinetic kernel modification is a constant mass term. However, there are very few renormalizable field theories where such a guess yields finite results, as emphasized by Coleman/1977/, and none of them in more than two dimensions. In a forthcoming publication, we shall treat the Gross-Neveu model in this way, and show that the procedure gives the exact answer for the vacuum energy density, using dimensional regularization and a three-loop calculation ignoring the usual trick of the introduction of the $\sigma = \overline{\psi}\psi$ field. This exact result coincides with the Hartree-Fock result, as was the case in the previous section for the anharmonic oscillator in lowest order.

At a given perturbative order, having found one solution of the variational equations for the kinetic kernel, it is quite interesting to expand the equations around this solution. In the example just mentioned of a constant solution, this expansion will involve non-trivial space-time dependence. We shall show that at first order in this expansion, the resulting linear equation is a lowest order Bethe-Salpeter equation, describing two-body bound states. Higher orders in the expansion generate non-linear interactions between these bound states, in the fashion of Callan, Coote and Gross/1976/ for two-dimensional QCD with N colors.

In another forthcoming paper, we shall also perform a non-trivial calculation in four-dimensional QCD.

Acknowledgements

I wish to thank Roger Dashen, whose remarks first made me contemplate the problem of improving perturbation theory fifteen years ago. I am grateful to Philippe Quentin for attracting my attention to the work of Roger Balian and Marcel Vénéroni, and to Bernard Bonnier, Gérard Mennessier and my other colleagues in Montpellier, for stimulating conversations.

References

Coleman S./1977/:*Classical lumps and their quantum descendents* Proceedings of the 1975 Erice summer school, A. Zichichi ed., Plenum Press, New-York.

Polley L. and D. Pottinger/1988/ editors, Proceedings of the Wangerooge conference on variational calculations in quantum field theories, World Scientific, Singapore.

Balian R. and Vénéroni M./1988/: Ann. Phys. (N.Y.) *187*, 29.
Duncan A. and Moshe M./1988/: Phys. Lett. *215B*, 352.
Jones H.F. and Monoyios M./1989/: Intern. J. Mod. Phys. *A4*, 1735, and references therein.
Seznec R. and Zinn-Justin J./1979/: J. Math. Phys. *20*,1398.
Callan C.G., Coote N. and Gross D.J./1976/: Phys. Rev. *D13*, 1694.

A New Method for QCD Sum Rules: The L^∞ Norm Approach

G. Mennessier, M.B. Causse, and G. Auberson

Laboratoire de Physique Mathématique*, Université Montpellier II, F-34095 Montpellier Cedex 05, France

Abstract

QCD is considered to be a good candidate for a theory of strong interactions. However, it is believed that non perturbative contributions are important at low energies, although their calculation from the QCD Lagrangian is a very difficult task. For a phenomenological approach, they have been parametrized in term of "condensates" [4]. Several methods to deal with these non perturbative contributions , the so-called QCD sum rules, have been used. Most of them do not take fully into account the errors in a quantitative and explicit way, and/or need assumptions on the derivatives of the amplitude. We argue that one should give oneself an a priori model of the errors on the amplitude itself, both in the time-like range and in the space-like range. A set of parameters (values of condensates or mass and width of a resonance) is then acceptable if there exists at least one analytic function going through the "error corridors". We develop approximate methods to solve this L^∞ norm problem.

1 Introduction

QCD [1] is widely considered to be a good candidate for a theory of strong interactions. Asymptotic freedom allows us to control the short distance behaviour by improving the perturbative expansion into a hopefully meaningful (asymptotic ?) running coupling expansion. Long distance behaviour and confinement are not fully understood, but non perturbative contributions, like instantons [2] and merons [3] are believed to be essential. However they are much more difficult to calculate. To make a comparison with experiment possible, even in the resonance energy range, S.V.Z. [4] have proposed to extend the use of the operator product expansion [5] into a parametrization of the non-perturbative effects. These contributions appear as a product of terms (calculable within perturbative theory) times the vacuum expectation values of local operators, the so-called condensates. The condensates incorporate the long distance effects and should be calculable from the Lagrangian. Only qualitative estimates have been possible up to now [6], and in a phenomenological approach, they are considered as parameters. It is thought that this parametrization provides us with a realistic model of amplitudes, for space-like values of the energy, down to the Gev range. Analyticity then correlates the values of the amplitude in the space-like domain, and in particular the values of the condensates, with its discontinuity in the time-like range, which is related to more physical and directly measurable quantities (to be more specific we have mainly in mind two-point functions of currents which have simple and well-known analyticity properties). In the absence of errors, the correlation would be very strong. However as soon as errors appear, it becomes much looser.

Several methods have already been devised to deal with this problem, based on various types of "sum rules" [4,7–19], or on several techniques of analytic extrapolation [20–25]. Most of the former include the "theoretical errors" in the space-like domain (or in the complex domain) only at a qualitative level, and/or need (explicit or implicit) assumptions on the derivatives of the amplitudes. But we know that, at large time-like values, there exists an infinite number of branch points due to the many particle thresholds, where only a finite number of derivatives of the true amplitude is finite. Thus, if it is reasonable to expect that the "QCD estimate" is a good "average value" of the amplitude (the old idea of duality), it is much more questionable (and in any case a much stronger physical assumption), to expect that it also provides us with a good estimate of the derivatives, even at high energies. Actually, two functions very close to one another may have quite different derivatives and inverse-Laplace or "Borel" transforms. The moment method [9] has already been criticized [24,26] for its instability when there exists a continuum under the resonance.

On the other hand, the approach based on analytic extrapolation is satisfactory from a mathematical point of view, but relies generally on the choice of some L^2 norms which seem physically rather arbitrary. Indeed, a chi-squared defined on a discrete set of points with a true statistical meaning could be satisfactory from a physical point of view, but in the present problem, one needs weight functions defined on a whole interval. These are devoided of a clear physical meaning, as it is manifest from the fact that they are not invariant against changes of the energy variable.

As a result of the inherent instabilities of some methods and of the poor way to deal quantitatively with the errors, the values found in the literature are often quite different and seem to be somewhat inconsistent, even if the problem does not suffer from deep instabilities (the determination of resonance parameters from QCD involves an extrapolation from interior to boundary and is unstable, while the determination of condensates from experimental data rather involves "interpolation" from the boundary and is mathematically stable). For instance, the values of the gluon condensate $< \alpha/\pi \; G^a_{\mu\nu} G^{\mu\nu}_a >$ obtained from the e^+e^- annihilation data vary from $.008 \pm .001 \mathrm{Gev}^4$ [20] or $.0115 \pm .00015$ [25] up to $.021 - .058$ [17]. Therefore it is very important to avoid any hidden assumption and to have a better control on the error analysis. We then propose to follow mainly the analytic extrapolation approach, which demands to make explicit the assumptions, and to introduce physically meaningful norms. Moreover we will use the idea of duality in a somewhat conservative way: QCD gives a good estimate of the amplitude, but small fluctuations with possibly large derivatives are allowed (for instance thresholds, weakly coupled but very narrow states,...).

As in most approaches, we need "experimental data" on the whole right hand cut $\Gamma = [s_0, \infty[$, which allows us to get an approximation of the imaginary part of the amplitude, let us call it f_0 , with some error channel σ_R. The "experimental data" f_0 can be theoretical estimates, models to be tested, or obtained from true experimental results by some fitting or interpolating method. We allow deviations from f_0 by at most $\pm\sigma_R$ at each point of Γ, and not only in average, which means in mathematical words that we are using an L^∞ norm.

In the space-like domain, on some interval $\gamma = [s_2, s_1]$(with possibly $s_2 = -\infty$), we get from QCD an approximate amplitude F_0 (the "space-like data") which may also depend on parameters as Λ, and as the values of some condensates. We insist on having also an a-priori estimate of the accuracy σ_L of the model F_0. This error function σ_L has to include errors due to the truncation of the perturbation expansion and of the non-perturbative contributions. It is expected to decrease at infinity where QCD is hoped to be good, to increase at lower values of $|s|$ and to diverge when QCD becomes completely meaningless.

We will say that F_0 and f_0 are *compatible* (or *consistent*) if there exists at least one function F, with required analytical properties, and with imaginary part f on the cut, such that:

$$|F(s) - F_0(s)| \leq \sigma_L(s) \; \forall s \in \gamma \tag{1.1}$$

and

$$|f(s) - f_0(s)| \leq \sigma_R(s) \; \forall s \in \Gamma. \tag{1.2}$$

Such a function is indeed a candidate to be the true amplitude. Given the "experimental data", we can then say that the set of values of the condensates is *acceptable*; or given the condensates, that, for example, the resonance parameters used to parametrize the right hand discontinuity are *allowed*.

If no analytic function satisfies the constraint of being inside the two error channels, then F_0 and f_0 are *inconsistent* and the set of associated parameters is *forbidden*.

Let us point out that the choice of σ_R and σ_L is a difficult and maybe controversial problem, but it is a physical one: a model or a theory without at least a rough estimate of its accuracy is useless!

In this article, we describe the mathematical methods used to answer this consistency question. An application of these methods to the determination of the allowed range of values for the condensates related to the annihilation $e^+e^- \to I = 1$ *hadrons* will be presented in a subsequent paper [27]. In section 2, we give the main theorems on the direct approach to this L^∞ norm problem, and we explain how to use them to answer the consistency question. The method may happen to be too difficult in some cases, and we describe in section 3 an indirect method based on a sequence of L^2 norm problems.

2 The L^∞ norm problem : direct approach

2.1 Introduction

Here, we restrict ourselves to the case of unsubtracted dispersion relations. The subtracted case will be discussed in paragraph 2.4. Specifically, the complex s-plane is cut along $\Gamma \equiv [s_0, \infty[$ with $s_0 > 0$, and the "space-like data" are given on the interval $\gamma \equiv [s_2, s_1]$ (with $s_2 < s_1 < s_0$ and possibly $s_2 = -\infty$).

There are given:

i) on γ : a continuous, real function F_0 and a continuous, strictly positive function σ_L.

ii) on Γ : a real (not necessarily continuous) function f_0 and a continuous function σ_R, strictly positive for $x > s_0$ ($\sigma_R(s_0) \neq 0$ is not required).

Functions with imaginary part f "not too far from f_0" have to satisfy an unsubtracted dispersion relation, thus we assume $f_0(x)/x$ and $\sigma_R(x)/x$ to be integrable :

$$f_0(x)/x \in L^1(\Gamma), \tag{2.1}$$

$$\sigma_R(x)/x \in L^1(\Gamma). \tag{2.2}$$

Let f be any function such that $(f - f_0)/\sigma_R$ is bounded:

$$\Delta(x) \equiv (f(x) - f_0(x))/\sigma_R(x) \in L^\infty(\Gamma). \tag{2.3}$$

Let $\mathcal{F}$ be the class of real analytic functions F of the form:

$$F(s) = \frac{1}{\pi}\int_\Gamma dx\ f(x)/(x - s), \tag{2.4}$$

and the function Ψ be defined by :

$$\Psi(y) \equiv (F(y) - F_0(y))/\sigma_L(y), \tag{2.5}$$

(Ψ measures the distance between F and the model F_0 in σ_L unit).

We now define the two following functionals over $\mathcal{F}$:

$$\chi_R[F] \equiv \|\Delta\|_\infty = essential \sup_{x\in\Gamma} |\Delta(x)|, \tag{2.6}$$

$$\chi_L[F] \equiv \|\Psi\|_\infty = \sup_{y\in\gamma} |\Psi(y)|. \tag{2.7}$$

Let $\mathcal{D}$ be the set of points (χ_R, χ_L) in R^2 such that there exists at least one function F in $\mathcal{F}$ with $\chi_R[F] \leq \chi_R$, $\chi_L[F] \leq \chi_L$. The physical problem is to determine wether $\mathcal{D}$ contains points $(\chi_R \leq 1, \chi_L \leq 1)$. Clearly, it is enough to construct the "lower boundary" $\partial\mathcal{D}$ of $\mathcal{D}$:

$$\chi_-(\chi) = \inf_{\|\Delta\|_\infty \leq \chi} \chi_L[F], \tag{2.8}$$

up to the point $\chi = 1$.

Introducing the function $\tilde{F}_0$ and the "discrepancy function" D_0 :

$$\tilde{F}_0(s) = \frac{1}{\pi}\int_\Gamma dx\ f_0(x)/(x-s), \tag{2.9}$$

$$D_0(y) = F_0(y) - \tilde{F}_0(y), \tag{2.10}$$

we can re-express Ψ as :

$$\Psi(y) = \frac{1}{\sigma_L(y)}(\frac{1}{\pi}\int_\Gamma \Delta(x)\sigma_R(x)/(x-y)\ dx - D_0(y)). \tag{2.11}$$

This allows us to consider the funtionals χ_R and χ_L over $\mathcal{F}$ as functionals over $L^\infty(\Gamma)$ and to restate the extremum problem (2.8) as :

$$\chi_-(\chi) = \inf_{\|\Delta\|_\infty \leq \chi} \|\Psi\|_\infty. \tag{2.12}$$

We notice that the problem involves only the function D_0, not F_0 and f_0 independently, and thus is rather stable with respect to local variations of f_0 (which only appears in D_0 through integration).

In 2.2, we describe the general properties of the function $\chi_-(\chi)$, and the structural properties of the extremal functions (i.e. of those functions F which turn out to saturate the bound χ_-). Then in 2.3 we show that these properties lead to a practical construction of the boundary $\partial\mathcal{D}$, at least in the case of a finite interval γ or for small enough χ_R. We make some extension and comments for the QCD case (subtraction and infinite γ) in 2.4. Since we need the boundary only up to $\chi_R = 1$ to get the answer about consistency, this construction will often be sufficient. However in some cases, we will need an other approach to solve the problem, which uses a sequence of L^2 norm problems, as explained in section 3.

2.2 Properties of the function χ_- and of the extremal functions

In ref.[28] it is proved that the infimum (2.8) or (2.12) is attained and that χ_- is a decreasing, convex and continuous function on (χ_0, ∞). In fact, $\chi_0 = 0$ for a bounded interval γ (finite s_2) or for an unbounded γ ($s_2 = \infty$) but a "reasonable" set of functions $F_0, f_0, \sigma_L, \sigma_R$. However when $s_2 = \infty$, χ_0 can be non zero and even infinite if the asymptotic behaviour of these functions is too badly chosen.

We need now to restrict ourselves to bounded γ. Only then, detailed information has been obtained on the extremal functions F_χ, i.e. on those functions which, given χ, saturate the bound $\chi_-(\chi)$ ($\chi_R[F_\chi] \leq \chi$, $\chi_L[F_\chi] = \chi_-(\chi)$ or equivalently $\|\Delta_\chi\|_\infty \leq \chi$, $\|\Psi_\chi\|_\infty = \chi_-(\chi)$ where Δ_χ, Ψ_χ are associated to F_χ through eq.(2.3) and (2.5)). Of course, we are also assuming that the extremal function F_χ does not identify with the function F_0 on γ, i.e. that $\chi_-(\chi) \neq 0$. Let us point out however that this (trivial) case may only occur if F_0 is itself the restriction of a function holomorphic in the complex plane cut along Γ, and if χ is large enough; in the QCD case this will not happen since the renormalization group improved amplitudes are singular already in the space-like domain, at $s = -\Lambda^2$.

In order to describe conveniently certain properties of F_χ, it is useful to introduce the number of its " *effective extrema* ". This essentially amounts to count the number of times the continuous function Ψ associated to a function F by eq.(2.5) reaches its extrema χ_- ($\equiv \chi_-(\chi)$) or $-\chi_-$ on γ, but counting only for 1 any set of *successive* extrema of the same sign (and not excluding the possibility that $\Psi(y) = \pm\chi_-$ on a whole subinterval of γ). The precise definition is given in [28] and an example is shown in Fig.1 .

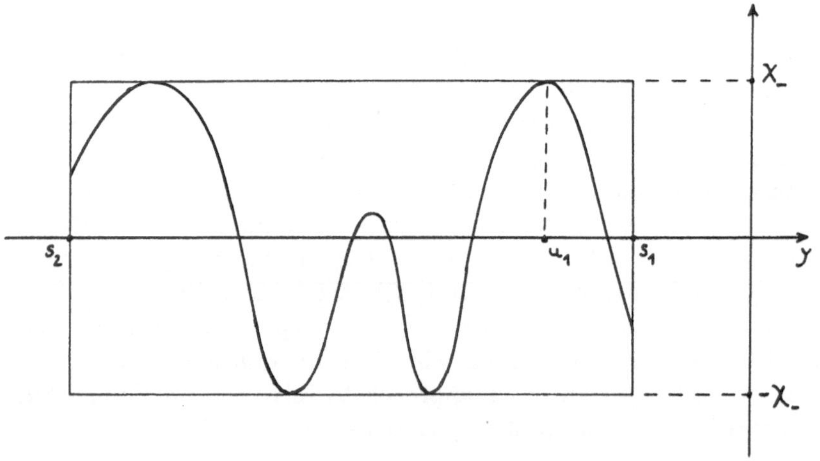

Fig.1: Typical graph of the function $\Psi(y)$ (full line). Here the number of effective extrema is $n = 3$ and $\varepsilon = 1$.

Let n be the number of effective extrema and ε the sign of the first extremum:

$$\varepsilon = \; sign\; of\; \Psi(u_1) = \pm 1, \tag{2.13}$$

where u_1 is the largest value of $y \leq s_1$ for which $|\Psi(y)| = \chi_-$. It is established that any extremal function Δ_χ obeys the condition:

$$|\Delta_\chi(x)| = \chi \;\; almost\; everywhere\; on\; \Gamma \tag{2.14}$$

and thus $\|\Delta_\chi\|_\infty = \chi$. In other words, the "absorptive part" of any extremal function Δ_χ saturates its bounds everywhere.

Thus it is meaningful to define *m-step functions*: a function defined on Γ is called a m-step function if it is right continuous, takes on the only values $\pm\chi$ and has $(m-1)$ jumps.

Then the following theorem holds: Let $\chi > 0$ be such that $\chi_-(\chi) \neq 0$ and F_χ be an extremal function with n effective extrema. Then its imaginary part $f_\chi(x) = Im F_\chi(x+i0)$ on Γ has the form:

$$f_\chi = f_0 + \sigma_R \Delta_\chi \tag{2.15}$$

where Δ_χ is a m-step function with $\|\Delta\| = \chi$. Moreover :

$$m \leq n \tag{2.16}$$

$$if\; m = n,\; then\; \varepsilon\Delta(s) < 0. \tag{2.17}$$

The uniqueness of the extremal function is then easily derived.

2.3 Practical construction of the boundary $\partial\mathcal{D}$

We will now show that the construction of the boundary reduces to a finite dimensional minimization, which can therefore be implemented on a computer (at least for small χ where the dimension turns out to be small).

Starting at $\chi = 0$, we have obviously $\Delta_\chi \equiv 0$ and from eq.(2.11) :

$$\Psi_0 = -D_0/\sigma_L. \tag{2.18}$$

It follows that

$$\chi_-(0) = \|\Psi_0\|_\infty = \varepsilon_0 \Psi_0(u_1) \tag{2.19}$$

is, in the generic case, strictly positive and is reached at only one point u_1, with $\varepsilon_0 = 1$. There is a finite gap between $\chi_-(0)$ and the absolute value of any other extremum with the opposite sign.

For non zero, but small enough values of χ, any allowed function F (with $\chi_R[F] = \|\Delta\|_\infty \leq \chi$) will give an associated function Ψ such that $|\Psi(y) - \Psi_0(y)|$ is smaller than half the gap on γ. Thus the number of effective extrema will still be 1 (see Fig.2). Because of the positivity of the Cauchy kernel, and in agreement with the theorem of section 2.2, the

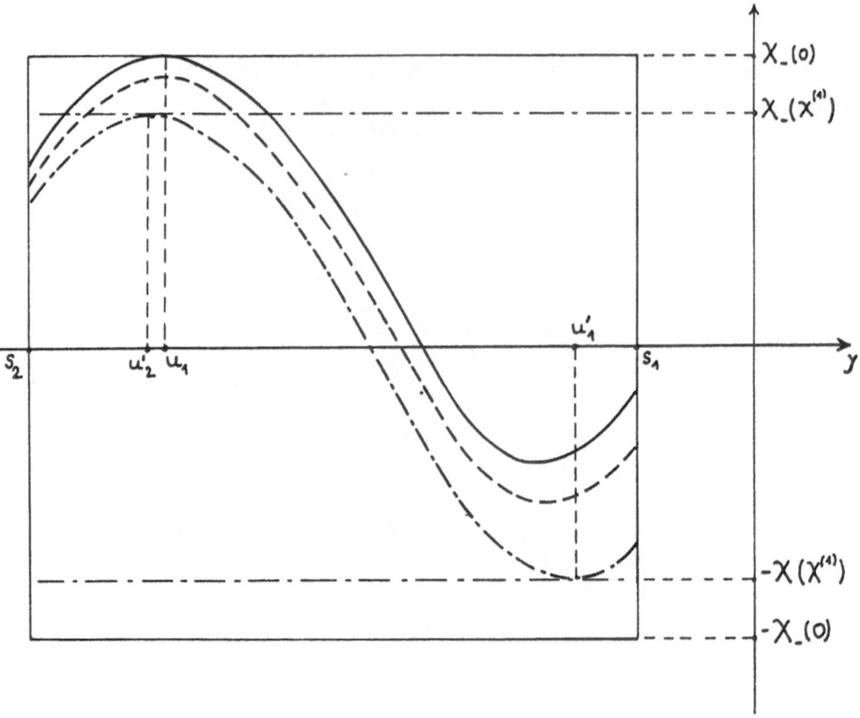

Fig.2: Generic graph of the extremal functions Ψ_χ for $0 \le \chi \le \chi^{(1)}$: Ψ_0 (full line), Ψ_χ (broken line), $\Psi_{\chi^{(1)}}$ (dot-dashed line).

extremal function is obtained by choosing the 1-step function $\Delta_\chi(x) = -\varepsilon_0\chi$ which decreases as much as possible the value of the supremum around u_1. Therefore, in that case, the extremal function is given by :

$$\Delta_\chi(x) = -\varepsilon_0\chi, \tag{2.20}$$

$$f_\chi = f_0 - \varepsilon_0\chi\sigma_R, \tag{2.21}$$

$$F_\chi(s) = \tilde{F}_0(s) - \varepsilon_0\chi\Sigma(s; s_0), \tag{2.22}$$

$$\Psi_\chi(y) = (-D_0(y) - \varepsilon_0\chi\Sigma(y; s_0))/\sigma_L(y), \tag{2.23}$$

where

$$\Sigma(s; x) = \frac{1}{\pi}\int_x^\infty dx'\ \sigma_R(x')/(x' - s). \tag{2.24}$$

One has just to compute the supremum of $|\Psi_\chi(y)|$ to get the minimum $\chi_-(\chi)$, which is very easy to do.

When χ increases, the value of the effective extremum of Ψ_χ decreases (in absolute value), but the absolute values of its (non effective) extrema with opposite sign increase. For a first critical value $\chi^{(1)}$, one of these extrema becomes an effective one. The extremal function Ψ_χ then gets two effective extrema, with abscissas $u' < u'$ (Fig.3). Thus for $\chi > \chi^{(1)}$, the function

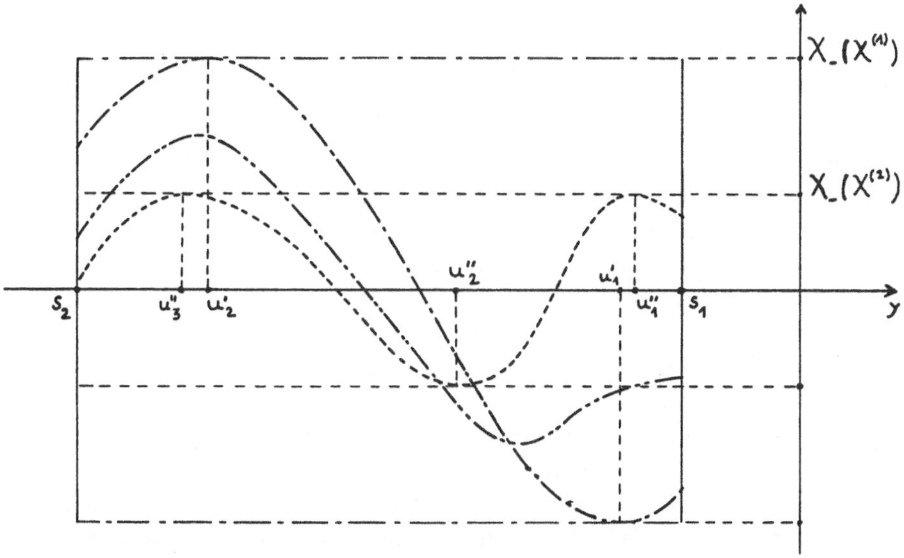

Fig.3: Generic graph of the extremal functions Ψ_χ for $\chi^{(1)} \leq \chi \leq \chi^{(2)}$: $\Psi_{\chi^{(1)}}$ (dot-dashed line), Ψ_χ (dot-dot-dashed line), $\Psi_{\chi^{(2)}}$ (dotted line)

given by (2.22) is no more extremal , and one must take a 2-steps function Δ_χ:

$$\Delta_\chi(x) = -\varepsilon_1\chi(\theta(x_1 - x) - \theta(x - x_1)) = -\varepsilon_1\chi(1 - 2\theta(x - x_1)) \qquad (2.25)$$

$$F_\chi(s) = \tilde{F}_0(s) - \varepsilon_1\chi(\Sigma(s; s_0) - 2\Sigma(s; x_1)) \qquad (2.26)$$

$$\Psi_\chi(y) = (-D_0(y) - \varepsilon_1\chi(\Sigma(y; s_0) - 2\Sigma(y; x_1)))/\sigma_L(y) \qquad (2.27)$$

where ε_1 is the sign of $\Psi_\chi(u')$ and x_1 is the jump abscissa. One has to minimize $\|\Psi\|_\infty$ as a function of x_1. Thus the problem is now a 1 parameter minimization. Moreover at the extremum, x_1 will be such that the maximum and the minimum of Ψ_χ will have opposite values, so that the problem is equivalent to a zero finding one, which is even easier. Let us point out that we have just one parameter x_1 to fulfil one constraint (minimum value opposite to maximum) which is enough in the generic case (and happens "accidentally" at the critical value $\chi^{(1)}$), and that we are not allowed to take more steps because of theorem (2.16).

As χ increases further, the maximum of Ψ_χ decreases while its minimum increases (they are equal in absolute value!). For a new critical value $\chi^{(2)}$, one local extremum becomes an effective one. The function Ψ_χ defined by eq.(2.27) then develops 3 effective extrema, with abscissas $u'' < u'' < u''$ (Fig.3). In the generic case, we then need a 3-step function Δ_χ for larger χ:

$$\Delta_\chi(x) = -\varepsilon_2\chi(\theta(x' - x) - \theta(x - x')\theta(x' - x) + \theta(x - x')) \qquad (2.28)$$

$$F_\chi(s) = \tilde{F}_0(s) - \varepsilon\chi(\Sigma(s;s_0) - 2\Sigma(s;x') + 2\Sigma(s;x')) \qquad (2.29)$$

and we have to deal with a 2 parameters minimization, namely the abscissas $x' < x'$ of the 2 jumps of Δ_χ, which is what we need to fulfil the 2 constraints (two equal extrema and one opposite).

For even larger values of χ, one encounters new critical values and one needs more and more steps. Except for the critical values where we have more effective extrema than steps, we have in general $n = m$. Though it seems to us unlikely, we have not been able to exclude another kind of critical value where one of the effective extrema would go out of γ through one of its end-points, leaving us with one less effective extrema, and allowing us to decrease the number of steps. However it is easy to prove that the number of steps n is unbounded when $\chi \to \infty$ (if $\chi_-(\chi)$ does not vanish for some finite χ). Indeed, let n be a fixed number of steps and $\rho(x_i)$ be the norm of the combination of analytic functions Σ appearing in equations analogous to eq.(2.26,29). The infimum ρ_n of $\rho(x_i)$ over the jump abscissas x_i is certainly strictly positive. For large enough χ, the norm of Ψ associated to any n-step function Δ, is at least of order $\chi\rho_n/max(\sigma_L)$. It would increase indefinitely with χ if n were bounded for the extremal functions Δ_χ, which contradicts the fact that $\chi_-(\chi) = \|\Psi_\chi\|_\infty$ is a decreasing function. Thus either we reach a finite value χ^{max} such that $\chi_-(\chi^{max}) = 0$, and the boundary $\partial\mathcal{D}$ is completely known (which means that F_0 was the restriction to γ of an holomorphic function in $C\backslash\Gamma$ with $\chi_R[F_0] = \chi^{max}$) or the number of steps will increase indefinitely with χ. However we need only to construct the boundary $\partial\mathcal{D}$ up to $\chi = 1$ or to a smaller value such that $\chi_- = 1$ (and even less using the convexity property), to decide about the consistency of the data. This often involves a small number of steps.

2.4 Remarks for certain QCD cases

In QCD cases of interest, there can be two new specific points: on the one hand the dispersion relation may need subtractions, and on the other hand, it is natural to assume that the model F_0 for the "space-like data" (the QCD amplitude) is known down to $s_2 = -\infty$, since this will add information in a domain where errors are thought to be small. The theorems giving characteristics of the extremal functions are then no more guaranteed to be valid, so that the previous construction may break down. Of course, physical results are expected not to be sensitive to what happens at very large energies (space or time-like), and it is proved indeed in [28] that the case $s_2 = -\infty$ can be recovered as the limit of $s_2 \to -\infty$. If an extremal function Ψ (on a bounded interval γ) has an extrapolation with absolute values near $-\infty$ smaller than its norm, it is also extremal on the infinite interval. However in the opposite case, this is no longer true, and one must be careful because the asymptotic behaviour of QCD amplitudes is often only logarithmic and can be very slow.

For definiteness, we will study the example of the two-point function of the vector hadronic current which needs one subtraction. Following [28], we assume, instead of eq.(2.1-2):

$$f_0(x)/x^2 \in L^1(\Gamma), \tag{2.30}$$

$$\sigma_R(x)/x^2 \in L^1(\Gamma). \tag{2.31}$$

The class $\mathcal{F}$ is now defined by:

$$F(s) = A_0 + \frac{s}{\pi}\int_\Gamma dx\ f(x)/(x(x-s)), \tag{2.32}$$

where A_0 is an arbitrary subtraction constant. The functions Δ and Ψ being still defined by eq.(2.3,5), the expression (2.11) is now replaced by:

$$\Psi(y) = \frac{1}{\sigma_L(y)}\Big(A_0 + \frac{y}{\pi}\int_\Gamma \Delta(x)\frac{\sigma_R(x)}{x(x-y)}dx + \tilde{F}_0(y) - F_0(y)\Big) \tag{2.33}$$

where

$$\tilde{F}_0(s) = \frac{s}{\pi}\int_\Gamma dx\ \frac{f_0(x)}{x(x-s)}, \tag{2.34}$$

and one has to minimize $\|\Psi\|_\infty$ over A_0 and Δ.

The general case has not been studied in detail. Since we have one parameter in addition to the jumps abscissas, we expect in the generic case to have one more extremum than steps. At small χ, we will start with 1 step and have just to minimize over A_0 which will leave room for 2 effective extrema at the optimum. Then we will have 2 steps and will have to minimize over A_0 and the jump abscissa....More generally we expect that in the generic case the number of effective extrema is equal to the number of subtraction constants plus the number of steps. This should apply for finite γ, or infinite γ when the asymptotic behaviour of the extremal functions is not too much constrained by the error function σ_L.

However, if σ_L is small enough, the subtraction constants can be determined from the asymptotic behaviour in the infinite γ case. For instance, let us come back to our precise example, and let us assume that at least the leading and next to leading terms have been calculated for F_0 and for the high energy behaviour of f_0. It is then natural to assume that the errors vanish at infinity as :

$$\sigma_R(x) \sim C\pi/Ln^{p+1}(x),\ \ C \neq 0 \tag{2.35}$$

$$\sigma_L(y) \sim C'/Ln^p(-y),\ \ C' \neq 0,\ p \geq 1 \tag{2.36}$$

and that $F_0(y)$ and $\tilde{F}_0(y)$ differ at infinity only by a constant F_{00}, which is completely determined by F_0 ,f_0 and a choice of the subtraction point. Thus the new discrepancy function :

$$\bar{D}_0(y) \equiv F_0(y) - \tilde{F}_0(y) - F_{00} \tag{2.37}$$

vanishes at infinity. Let its behaviour be:

$$\bar{D}_0(y) \sim C''/Ln^p(-y) \tag{2.38}$$

where $C'' = 0$ if the vanishing is faster. Since $\sigma_R(x)/x$ is integrable (in contradistinction to f_0) eq.(2.32) can be written as:

$$\Psi(y) = \frac{1}{\sigma_L(y)}\Big(\frac{1}{\pi}\int_\Gamma \Delta(x)\frac{\sigma_R(x)}{(x-y)}dx - \bar{D}_0(y) + (A_0 - F_{00} - \frac{1}{\pi}\int_\Gamma dx\, \Delta(x)\frac{\sigma_R(x)}{x})\Big) \tag{2.39}$$

In the right hand bracket of eq.(2.39), the first two terms vanish at $y = -\infty$, while the third one is a constant. On an infinite interval γ, since σ_L vanishes at infinity, one must set this term to zero (otherwise the norm of Ψ will be infinite). Therefore the "best" A_0 is:

$$A_0 = F_{00} + \frac{1}{\pi}\int_\Gamma dx\, \Delta(x)\, \sigma_R(x)/x \tag{2.40}$$

The minimization over Δ is then exactly the same problem as in the unsubtracted case, and is defined by (2.11,12) with $\bar{D}_0$ substituted for D_0, and with an infinite interval γ.

Starting again at $\chi = 0$, we observe that the function Ψ_0 has only one effective extremum in the generic case. When C'' is not too large, this extremum is attained at a finite value of y, and there exists a finite gap between $|\Psi_0(y)|$ and $\|\Psi\|$ for any large enough $|y|$. As long as this property holds, the minimization of Ψ over an infinite γ or over a large but finite sub-interval are equivalent, and the previous construction applies.

However, the critical values $\chi^{(i)}$ now have an accumulation point at a finite value $\chi^{(\infty)}$. Indeed, the functions Σ involved in the expression of the extremal functions (2.26,29) satisfy:

$$\Sigma(y;x)/\sigma_L(y) \to C''' = C/(pC') \tag{2.41}$$

while the contribution to Ψ of the dispersion integral over any finite part of Γ behaves as $1/y\sigma_L(y)$ and vanishes at infinity. Thus, the asymptotic behaviour of a function Ψ associated to a function Δ with a finite number of steps is governed only by its last, infinite step and is

$$\pm\chi C''' - C'' . \tag{2.42}$$

This increases (in absolute value) with χ, and would be a lower bound for χ_- if the extremal functions Δ_χ had only a finite number of steps, which, above some value of χ, would contradict the decrease of $\chi_-(\chi)$. This shows that for χ larger than some $\chi^{(\infty)}$ the previous characterization of extremal functions is no more valid: Ψ_χ has an infinite number of effective extrema and Δ_χ is likely to have an infinite number of steps (In fact one has to modify a little bit the definition of an effective extremum in order to take into account the possible existence of sequences of alternate local extrema, strictly different from $\|\Psi\|_\infty$ but with values accumulating at $\pm\|\Psi\|_\infty$, when $y \to -\infty$).

In the next section, we describe another approach, involving a sequence of L^2 norm problems. It will allow us to get lower and upper bounds on the boundary of $\mathcal{D}$, which is useful when there are many steps (or an infinite number of them).

3 The L^2 norm sequence approach

3.1 Introduction

Being given the same data functions as in section 2, and the same class of analytic functions $\mathcal{F}$, we now define L^2 norms in addition to the L^∞ norms (2.6,7). For this, we need a set of two arbitrary weight functions $w = (w_R, w_L)$. More precisely, let w_R be a normalized positive function,

$$\int_\Gamma dx\ w_R(x) = 1\ , w_R(x) \geq 0, \tag{3.1}$$

such that:

$$\sigma_R(x)/(x w_R(x)) \in L^\infty(\Gamma), \tag{3.2}$$

(which enforces $w_R(x)$ to be **strictly** positive for $x > s_0$, from the properties of σ_R); and let w_L be a positive, normalized function:

$$\int_\gamma dy\ w_L(y) = 1\ , w_L(y) \geq 0. \tag{3.3}$$

We define the following functionals over $\mathcal{F}$:

$$\chi_{wR}[F] = \|\Delta\|_w \equiv (\int_\Gamma dx\ w_R(x)|\Delta(x)|^2)^{1/2}, \tag{3.4}$$

$$\chi_{wL}[F] = \|\Psi\|_w \equiv (\int_\gamma dy\ w_L(y)|\Psi(y)|^2)^{1/2}. \tag{3.5}$$

which, as in section 2, can be considered as functionals over the functions $\Delta \in L^\infty(\Gamma)$.

Let $\mathcal{D}_w$ be the set of points (χ_R, χ_L) in R^2 such that there exists at least one F with $\chi_{wR}[F] \leq \chi_R$, $\chi_{wL}[F] \leq \chi_L$, and let $\partial\mathcal{D}_w$ be its "lower boundary":

$$\chi_{w-}(\chi) = \inf_{\chi_{wR}[F]\leq\chi} \chi_{wL}[F] = \inf_{\|\Delta\|_w\leq\chi} \|\Psi\|_w. \tag{3.6}$$

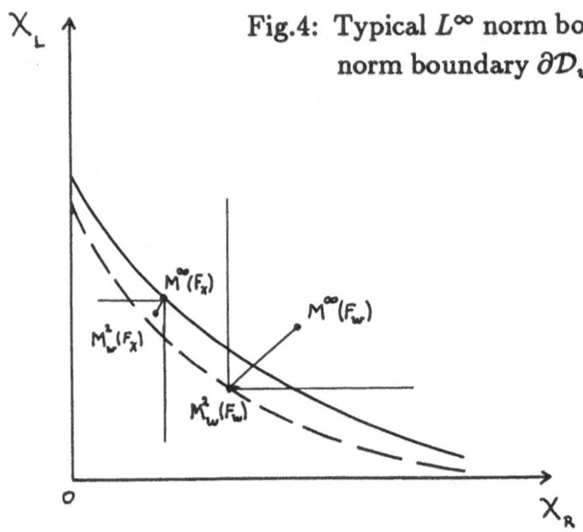

Fig.4: Typical L^∞ norm boundary $\partial\mathcal{D}$ (full line) and L^2 norm boundary $\partial\mathcal{D}_w$ (broken line)

For any choice of the couple w of weight functions and any function F in $\mathcal{F}$ we obviously have :

$$\chi_{wR}[F] \leq \chi_R[F] \tag{3.7}$$

and

$$\chi_{wL}[F] \leq \chi_L[F]. \tag{3.8}$$

Because $\chi_{w-}(\chi)$ is a decreasing function, this immediately shows that the boundary $\partial\mathcal{D}_w$ is "below" $\partial\mathcal{D}$ for any w. Let us define the points in R^2, $M^\infty(F) = (\chi_R[F], \chi_L[F])$, and $M^2(F) = (\chi_{wR}[F], \chi_{wL}[F])$. Then for any extremal function F_w of the L^2 norm problem, we have (Fig.4):

$$M^2(F_w) \in \partial\mathcal{D}_w \text{ "below" } \partial\mathcal{D},\ \partial\mathcal{D} \text{ "below" } M^\infty(F_w) \tag{3.9}$$

which allows us to derive "lower and upper" bounds on $\partial\mathcal{D}$ if we are able to get the extremal functions of the L^2 norm problem and to compute their associated L^∞ and L^2 norms. But this is a usual and well known problem (see for instance ref.[22,29]).

Moreover, we will argue that we can compute a sequence of functions $w^n = (w^n_R, w^n_L)$ which is likely to improve these bounds. The convergence toward points of $\partial\mathcal{D}$ has not been proved,but actual calculations lead to good bounds, allowing us to construct the L^∞ norm boundary $\partial\mathcal{D}$ with the convenient accuracy in the useful range.

3.2 Properties of the functions χ_{w-} and determination of the extremal functions

Using the same methods as for the L^∞ norm problem, one can first prove that the infimum (3.6) is attained and that $\chi_{w-}(\chi)$ is a decreasing, convex and continuous function on some interval (χ_0', ∞) (with $\chi_0' \leq \chi_0$ as defined in 2.2). Then, following the usual method of Lagrange multipliers, we define the Lagrangian:

$$\mathcal{L}(F) = \sin\theta \int_\gamma dy\ |\Psi(y)|^2 w_L(y) + \cos\theta \int_\Gamma dx\ |\Delta(x)|^2 w_R(x). \tag{3.10}$$

This expression has a simple geometric interpretation in the (χ_R^2, χ_L^2) plane. It is the distance to the origin of a straight line with slope $\tan(\theta + \pi/2)$ passing through the point $(\chi_R^2[F], \chi_L^2[F])$. When $\mathcal{L}$ is minimal, this gives a tangent to $\partial\mathcal{D}_w$ in the variables χ_R^2, χ_L^2 so that the boundary appears as a convex envelope of straight lines when θ is varying on $(0, \pi/2)$ (in the variables χ_{wR}, χ_{wL}, this corresponds to an envelope of ellipses) .

Since $\mathcal{L}$ is quadratic in the unknown function Δ, we obtain a linear (Fredholm) equation at the extremum:

$$\sin\theta\ \sigma_R(x)\Big(\frac{1}{\pi}\int_\Gamma dx' K(x,x')\sigma_R(x')\Delta(x') + h(x)\Big) = -\cos\theta\ w_R(x)\Delta(x) \tag{3.11}$$

with

$$K(x,x') = \frac{1}{\pi}\int_\gamma dy\ \frac{w_L(y)}{\sigma_L^2(y)} \frac{1}{(y-x)(y-x')} \tag{3.12}$$

$$h(x) = \frac{1}{\pi}\int_\gamma dy\ \frac{w_L(y)}{\sigma_L^2(y)} \frac{D_0(y)}{(y-x)}\ . \tag{3.13}$$

The conditions (2.2),(3.1-3) and compactness of γ (s_2 finite) or of the support of w_L, then imply that the symmetrized kernel is Hilbert-Schmidt and negative (for $0 \leq \theta < \pi/2$) and that the corresponding inhomogeneous term is squared integrable. Thus the solution is well defined for given θ and w, and (3.2) implies that Δ is bounded (see also [29] for detailed results on quite similar L^2 norm problems). The threshold and asymptotic behaviours of Δ turn out to be:

$$\Delta(x) \sim_{x\to s_0} C^t\sigma_R(x)/w_R(x) \tag{3.14}$$

$$\Delta(x) \sim_{x\to\infty} C^t\sigma_R(x)/(x w_R(x)). \tag{3.15}$$

The extremal function $F_{w\theta}$ is then easily computed from the dispersion integral, and one gets both points $M^\infty(F_{w\theta})$ and $M^2(F_{w\theta})$ which give lower and upper bounds for $\partial\mathcal{D}$. Moreover, we know that, in the χ^2 variables, the boundary is above the straight line going through the point $M^2(F_{w\theta})$ with slope $\tan(\theta + \pi/2)$.

3.3 Choice of the weight functions

We have to choose couples w of functions in such a way that the solution of the L^2 norm problem be as close as possible to that of the L^∞ norm problem. In particular the solution Δ of (3.11) should approach a step-like function. Thus we shall impose:

$$w_R(x) \sim_{x\to s_0} C^t \sigma_R(x), \tag{3.16}$$

$$w_R(x) \sim_{x\to\infty} C^t \sigma_R(x)/x, \tag{3.17}$$

which is compatible with the constraint (3.2) and, by eq. (3.14,15), will give a constant behaviour of Δ at least near s_0 and infinity.

For a fixed θ, let us assume that we have solved (3.11) for the couple $w^n=(w_R^n, w_L^n)$. The question is now : how to define a new couple w^{n+1} such that the solution of the new L^2 norm problem is likely to provide us with better bounds?

Let us begin by improving only w_R (and assuming $w_L^{n+1} = w_L^n$). In the left hand side of (3.11), Δ only appears through integration. Thus one can expect that the results of variations of Δ will be smoothed there. We would like to get $\Delta^{n+1}(x) \simeq \pm$ *some constant* χ, step-wise. Therefore, neglecting the small variations of the left-hand side when going from n to $n+1$, we should have:

$$left-hand\ side = -\cos\theta\ w_R^n\Delta^n \simeq -\cos\theta\ w_R^{n+1}(\pm\chi). \tag{3.18}$$

This suggests the choice :

$$w_R^{n+1}(x) \simeq w_R^n(x)|\Delta^n(x)|/\|\Delta^n\|_w. \tag{3.19}$$

Indeed, the iteration (3.19) will produce an increase (resp. a decrease) of the weight $w_R^{n+1}(x)$ at points where $|\Delta^n(x)|$ is large (resp. small). In this way, $|\Delta^{n+1}(x)|$ will tend to increase relative to $|\Delta^n(x)|$ wherever the latter function is small, and one can hope to obtain a better approximation of a step-function. This is true as long as there are not too many cancellations in the left-hand side of (3.11). If this l.h.s. vanishes at some point ξ, Δ^n vanishes at this point too, as well as w^{n+1}, which is forbidden by condition (3.2) (the eq.(3.11) is no more guaranteed to have a good Hilbert-Schmidt kernel). Moreover, one would have $w_R^{n+1} \simeq C^{te}|x-\xi|$, and the left hand side would vanish at the $n+1^{th}$ iteration at a point $\xi+\eta$ (with η small but non zero in general). According to eq.(3.11), the function Δ^{n+1} near $x=\xi$ would then behave roughly as:

$$\Delta^{n+1}(x) \simeq C^{te}(x-\xi-\eta)/|x-\xi|. \tag{3.20}$$

This has indeed a step-like behaviour as soon as $|x-\xi|$ is large as compared to η, but has a strong oscillation at $x=\xi$, on a scale $|\eta|$. Therefore we need to smooth w_R^{n+1} around ξ in order to avoid such a phenomenon (for instance through a convolution with neighbouring points). Let us remind that we wish to get not only a small L^2 norm (which is not sensitive to oscillations on a very small interval), but also a small L^∞ norm for Δ, in order to obtain good upper bounds on $\partial\mathcal{D}$ (these two norms coincide only for step-functions if w_R is non zero everywhere). In practical computations, an appropriate choice is (up to a normalization factor):

$$w_R^{n+1}(x) \simeq w_R^n(x)(C_1 + |\Delta^n(x)|/\|\Delta^n\|_w) \tag{3.21}$$

with C_1 of order unity (to avoid too fast variations of Δ), and with a smoothing of w_R around the zeroes it tends to develop. In the $n \to \infty$ limit, w_R should vanish exactly at the points where the solution Δ of the L^∞ norm problem have jumps, although such a limiting w_R is outside the class of allowed weight functions. Clearly, the convergence toward the desired solution may be achieved only through carefully chosen sequences of weight functions, which is likely to make difficult any rigorous proof of convergence.

As for the choice of w_L, we also wish its L^2 norm to approach its L^∞ norm. However we know that the extremal functions Ψ of the L^∞ norm problem are continuous and, in the generic case, that they attain their extremal values only at some isolated points. Thus we need to concentrate the weight just around these points, otherwise the L^2 norm will be much smaller than the L^∞ norm (and will give a "bad" lower bound on $\partial\mathcal{D}$). In the n infinite limit, w_L should tend towards a sum of δ-functions at these points. Of course, we do not know at the begining where they are, but, when iterating, we can increase the weight $w_L(y)$ at points where $|\Psi^n(y)|$ is large (and presumably not too far from its extrema) and decrease it everywhere else. Many choices are possible. In practical computations we have taken (up to a normalization):

$$w_L^{n+1}(y) \simeq w_L^n(y)(C_2 + |\Psi^n(y)|^2/\|\Psi^n\|^2) \tag{3.22}$$

with C_2 of order 0.1 . We do not apply eq.(3.22) at every iteration, but only every 2 to 5 iterations to allow for a stabilization of w_R after each change of w_L. Then we compute such iterations for each value of θ needed to get $\partial\mathcal{D}_w$ with the required accuracy.

3.4 Remarks for certain QCD cases

For definiteness, let us assume that we are in the case described in 2.3, so that we can define the function $\bar{D}_0$. Choosing the subtraction constant A_0 as in section 2.3 (which may not be the best L^2 norm choice !), we are left with the same minimization problem as in section 3.3, with D_0 replaced by $\bar{D}_0$.

Thus, the main difference is again that γ extends down to $s_2 = -\infty$. For not too large values χ of the norm of Δ, we know that the extremal functions of the L^∞ norm problem have only a finite number of effective extrema, so that we need the weight functions w_L^n to be non-zero only over a finite range. Thus, below some value y^c, it is advisable to choose $w_L^n(y) = 0$, and to check afterwards that the associated Ψ does not develop larger extrema for $y < y^c$.

However, as explained in section 2.4, in a case of logarithmic asymptotics, the behaviour of Ψ at $y \to -\infty$ is driven by that of Δ at $x \to +\infty$. If we choose w_R in such a way that the L^2 norm extremal solution Δ approaches χ (or$-\chi$) at $x \to \infty$, its associated Ψ will behave as $\chi\ C''' - C''$: even if the L^2 norms of Ψ and Δ give a good lower bound, their L^∞ norm will be rather large (for large χ) and will give an uninteresting upper bound. Therefore, when χ is large (i.e. $\chi\ C''' - C'' > \chi_-(\chi)$), in order to improve the behaviour of Ψ at infinity without changing it too much at finite energy, we increase the weight w_R^{n+1} of formula (3.21) by a factor of order $C'''\|\Delta^n\|_w/\|\Psi^n\|_w$, above some large value x_m. It is then expected that the corrresponding solution Δ of eq.(3.11) will be of order less than $\|\Psi\|_w/C'''$ for $x > x_m$ and that the supremum of $\|\Psi\|$ will not be driven by its behaviour at infinity. In actual computations, we thus have chosen in such cases the iteration rule :

$$w_R^{n+1}(x) \simeq w_R^n(x)(C_1 + C_3|\Delta^n(x)|/\|\Psi^n\|_w) \tag{3.23}$$

instead of eq.(3.21) for $x > x_m$,with $C_3 \simeq 2C'''$. In Fig.5 is shown an example of "bounds" for the curve $\chi_-(\chi)$ obtained by this iteration method.

4 Conclusion

The large spread of results for the values of the condensates found in various "QCD sum rules" analysis of the same data gives evidence that one must improve the methods and control the errors in a better way. Some usual methods are also affected by potential instabilities and have already been critized. A method must be well defined from a mathematical point of view and physically meaningful. We have argued that one must give one-self explicitly a quantitative estimate of the errors including both experimental and theoretical ones (truncation of the perturbative and operator product expansions). Moreover we think (conservatively?) that it is easier and more physical to estimate the errors for the amplitude itself rather than for some derivatives or Laplace transform. Then we say that a set of parameters is *admissible* (or *compatible*, or *consistent*) if there exists at least one function with the required analytical properties which goes through the error corridors (in a point-wise sense, not only in χ^2 average). We have shown in section 2 that answering this question can be reduced to a minimization over a finite number of variables (at least for a finite interval γ and , in the "good" cases, for an infinite γ), and to a sequence of Fredholm equations in the worse cases (section 3), which leads to practicable computations.

Such an approach emphasizes the importance of error estimates. Of course, the choice of these estimates is not always easy and may be controversial, but it is a physical problem. An application to the determination of

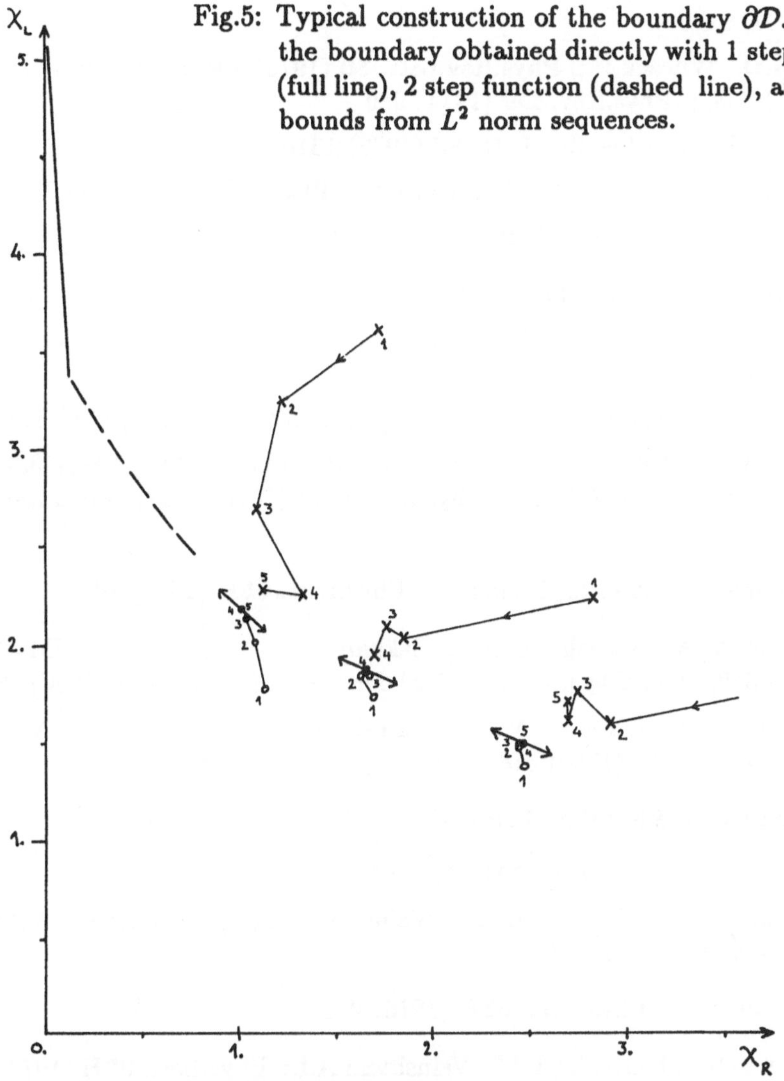

Fig.5: Typical construction of the boundary $\partial\mathcal{D}$. Part of the boundary obtained directly with 1 step function (full line), 2 step function (dashed line), and bounds from L^2 norm sequences.

condensates from e^+e^- annihilation into hadrons data has been completed, and will be presented in a subsequent paper [27].

Finally, let us notice that these ideas can apply to other kernels such as the "Borel-Laplace" ones, and could be interesting if it were to turn out that the inverse Laplace transforms (say) of certain Green's functions were better understood and controlled from a theoretical point of view than these Green's functions themselves.

Acknowledgements

We thank S.Ciulli and T.D.Spearman for useful discussions.

References

[1] Gross D., Wilczek F.; Phys.Rev.Lett. **30** (1973) 1343; Phys.Rev. **D8** (1973) 3633; Phys.Rev. **D9** (1974) 980

Politzer H.D.; Phys.Rev.Lett. **30** (1973) 1346

Fritzch H., Gell-Mann M., Leutwyler H.; Phys.Lett. **47B** (1973) 365

Weinberg S.; Phys.Rev.Lett. **31** (1973) 494

[2] Polyakov A.M.; Phys.Lett. **59B** (1975) 82

Belavin A.A., Polyakov A.M., Schwartz A.S., Tyupkin Y.S.; Phys.Lett. **59B** (1975) 85

't Hooft G.; Phys.Rev.Lett. **37** (1976) 8; Phys.Rev. **D14** (1976) 3432

Callan C.G., Dashen R., Gross D.J.; Phys.Lett. **63B** (1976) 334; Phys.Lett. **66B** (1977) 334; Phys.Rev. **D17** (1978) 2717; Phys.Rev. **D19** (1978) 1826

[3] de Alfaro V., Fubini S., Furlan G.; Phys.Lett. **65B** (1977) 163

[4] Shifman M.A., Vainshtein A.I., Zakharov V.I.; Nucl.Phys. **B147** (1979) 385; Nucl.Phys. **B147** (1979) 448; Nucl.Phys. **B165** (1980) 45

Novikov V.A.,Shifman M.A., Vainshtein A.I., Zakharov V.I.; Nucl.Phys. **B174** (1980) 378; Nucl.Phys. **B191** (1981) 301

[5] Wilson K.G.; Phys.Rev. **179** (1969) 1499

[6] Shuryak E.V.; Phys.Rep. **115** (1984) 151

Novikov V.A.,Shifman M.A., Vainshtein A.I., Zakharov V.I.; Phys.Rep. **116** (1984) 103

[7] Yndurain F.J.; Phys.Lett. **63B** (1976) 211

[8] Eidelman S.I., Kurdadze L.M., Vainshtein A.I.; Phys.Lett. **82B** (1979) 278

[9] Bell J.S., Bertlmann R.A.; Z.Phys. **C4** (1980) 11

[10] Reinders L.J., Rubinstein H.R., Yazaki S.; Phys.Lett. **95B** (1980) 103; Phys.Rep. **127** (1985) 1

[11] Narison S.,de Rafael E.; Phys.Lett. **103B** (1981) 57

[12] Krasnikov N.V., Pivovarov A.A., Tavkhelidze N.N.; Z.Phys. **C19** (1983) 301

[13] Launer G., Narison S., Tarrach R.; Z.Phys. **C26** (1984) 433

[14] Bertlmann R.A., Launer G., de Rafael E.; Nucl.Phys. **B250** (1985) 61

[15] Dominguez C.A.; Proc.International Workshop on Quarks,Gluons and Hadronic Matter (University of Cape Town 1987), DESY preprint 87-002

[16] Fischer J., Kolar P.; Z.Phys. **C34** (1987) 375

[17] Bertlmann R.A., Dominguez C.A., Loewe M., Perrottet M., de Rafael E.; Z.Phys. **C39** (1988) 231

[18] Nasrallah N.F., Papadopoulos N.A., Schilcher K.; Phys.Lett. **113B** (1982) 61; Z.Phys. **C16** (1983) 323

[19] Kremer M., Nasrallah N.F., Papadopoulos N.A., Schilcher K.; Phys.Rev. **D34** (1986) 2127

[20] Caprini I., Verzegnassi C.; Nuovo Cim. **75A** (1983) 275; Nuovo Cim. **90A** (1985) 388

[21] Caprini I.; Proc.of International Symposium on Hadron Interactions (Bechyne 1988, Czechoslovakia) Bucharest preprint

[22] Ciulli M., Ciulli S., Spearman T.D.; J.Math.Phys. **25** (1984) 3194

[23] Stefanescu I.S.; J.Math.Phys. **27** (1986) 2657

[24] Ciulli S., Geniet F., Mennessier G., Spearman T.D.; Phys.Rev. **D36** (1987) 3494

[25] Misra S., Deo B.B.; Phys.Rev. **D34** (1986) 254; Phys.Rev. **D37** (1988) 1325

[26] Bowcock J., Ciulli S., Geniet F.; Montpellier preprint
Geniet F.; Thesis, University of Montpellier, U.S.T.L. (1987)

[27] Causse M.B., Mennessier G.; Z.Phys. **C47** (1990) 611

[28] Auberson G., Mennessier G.; Comm.Math.Phys. **121** (1989) 49

[29] Ciulli M.; Ph.D. Thesis, School of Math. University of Dublin, Trinity College (1988)

Bifurcation Theory
Applied to Chiral Symmetry Breaking

D. Atkinson

Institute for Theoretical Physics, R.U.G., P.O. Box 800,
9700 AV Groningen, The Netherlands

Abstract
Chiral symmetry breaking in quantum electrodynamics and quantum chromodynamics is considered as a problem in bifurcation theory. Inequalities and positivity play key rôles, as they do in much of the work of André Martin.

1 Chiral Symmetry Breaking

In the absence of a bare fermion mass, the Lagrangians of QED and QCD are chirally symmetric: the left and right chiral projections decouple from one another. This situation persists to all orders of perturbation theory: the left hand is not perturbed by what the right hand does.

Beyond perturbation theory, however, the Dyson-Schwinger equation for the fermion mass-function can have nontrivial solutions, if the coupling is large enough. In ordinary QED, this does not happen, for the fine structure constant is too small; but a strong-coupling phase may exist. In QCD, the running coupling constant ensures ultra-violet asymptotic freedom; but it is the low-energy value of the coupling that is relevant to chiral symmetry breaking. It is thought that the effective infra-red coupling is infinite (confinement), which then guarantees, but is not required by, the non-perturbative breaking of chiral symmetry and the generation of a quark mass.

2 QED

In the Landau gauge, in ladder approximation, the electron mass-function, $m(p^2)$, satisfies the integral equation

$$m(x) = \lambda \int_0^{\Lambda^2} \frac{dy}{\max(x,y)} \frac{y m(y)}{y + m^2(y)}, \tag{1}$$

where $x = p^2$ is the square of the momentum, in Euclidean space, λ is proportional to the square of the electron charge, and Λ^2 is an ultra-violet cut-off, necessary for the existence of a critical point, so long as the effective charge is not allowed to run.

We wish to show that (1) has only the trivial solution if $\lambda < \lambda_c(\Lambda)$: it does have a nontrivial solution when λ is greater than this bifurcation point, and moreover $\lambda_c(\Lambda)$ has a strictly positive limit as $\Lambda \to \infty$. We first consider the *quasilinear* auxiliary equation,

$$m_0(x) = \lambda \int_0^{\Lambda^2} \frac{dy}{\max(x,y)} \frac{y m_0(y)}{y + m_0^2}, \tag{2}$$

with the normalization

$$m_0(0) = m_0. \tag{3}$$

The integral equation (2) may be reduced to a hypergeometric differential equation, with infra-red and ultra-violet boundary conditions. The solution is

$$m_0(x) = m_0 F\left(\frac{1}{2} + \mu, \frac{1}{2} - \mu; 2; -\frac{x}{m_0^2}\right), \tag{4}$$

with the ultra-violet condition

$$B(\Lambda^2) = 0, \tag{5}$$

where

$$B(x) = m_0 F\left(\frac{1}{2} + \mu, \frac{1}{2} - \mu; 1; -\frac{x}{m_0^2}\right). \tag{6}$$

In the above formulas,

$$\mu = \left(\frac{1}{4} - \lambda\right)^{\frac{1}{2}}; \tag{7}$$

and it can be shown readily, from the properties of the hypergeometric function, that, for $0 < \lambda < \frac{1}{4}$, i.e. μ real, both $m_0(x)$ and $B(x)$ are real and positive for all $x > 0$. Hence (5) cannot be satisfied for *any* finite Λ. Consequently there is no nontrivial solution of (2), if $\lambda < \frac{1}{4}$ and $\Lambda < \infty$. For $\lambda > \frac{1}{4}$, on the other hand, μ is imaginary, $B(x)$ oscillates, and (5) can always be met by choosing m_0 suitably.

It should be stressed that, without the U-V cut off, Eq. (2) has a nontrivial solution for any real λ: it is the introduction of the cut-off, or equivalently the requirement that $B(x)$ oscillate, that forces the critical point, λ_c, to be greater than 0.25.

The nonlinear equation (1) has been analyzed /Atkinson, 1988a/ by comparing a solution, $m(x)$, with a solution of (2), both with the same value of $m(0) = m_0$. The condition that a solution of the differential equation,

$$\frac{d}{dx}\left[x^2 \frac{dm}{dx}\right] + \lambda x \frac{m}{x + m^2} = 0 \tag{8}$$

satisfy (1) is that

$$A(x) = m_0 - \lambda \int_0^x dy \frac{m(y)}{y + m^2(y)} \tag{9}$$

have a zero at the UV cut-off point, $x = \Lambda^2$. In the domain of the positivity of $m(x)$, namely

$$\mathcal{D}(m) = \{x | x > 0; \quad m(y) \geq 0, \quad y \in [0, x]\}, \tag{10}$$

it can be shown that $m_0 \geq m(x) \geq A(x)$, $m(x) \leq m_0(x) \leq m_0$ and

$A(x) \leq B(x) \leq m_0$, so it follows that, if $B(x)$ has a zero at $x = x_1$, then $A(x)$ has a zero at a point $x_0 < x_1$, and from this it is easy to see that (1) always has a nontrivial solution if $\lambda > \lambda_c(\Lambda) > \frac{1}{4}$, just as in the quasilinear case (2).

3 QCD

In a nonabelian gauge theory like QCD, the logarithmic decrease of the effective coupling in the UV effectively dispenses with the need for a cut-off. Instead of (1), we now have

$$m(x) = \lambda \int_0^\infty \frac{dy}{\max(x,y)\log[\tau + \max(x,y)]} \frac{ym(y)}{y + m^2(y)}, \tag{11}$$

and its quasilinear approximant,

$$m_0(x) = \lambda \int_0^\infty \frac{dy}{\max(x,y)\log[\tau + \max(x,y)]} \frac{ym_0(y)}{y + m_0^2}, \tag{12}$$

again with the normalization (3). Here τ is a parameter greater than unity that acts as an infra-red cut-off.

It is easy to determine the UV behaviour of the most general solution of (11):

$$m(x) \sim \frac{C}{x+\tau} \log^{-1+\lambda}(x+\tau) + B \log^{-\lambda}(x+\tau), \tag{13}$$

for $x \to \infty$, and similarly for (12). There is in fact a nontrivial solution for any value of λ; but it has been shown /Miransky,1985/ that, in the absence of explicit chiral symmetry breaking, one must impose $B = 0$. An alternative formulation of the same requirement is that $m(x)$ be square integrable.

For a fixed value of m_0, the quasilinear equation (12) is a homogeneous Fredholm equation, and so it has a nontrivial L^2 solution only if λ belongs to a point set. Let λ_0 be the smallest such eigenvalue. As $m_0 \to 0$, it might naïvely be expected that $\lambda_0 \to 0$; but this is not true, for it has been shown numerically /Atkinson, 1988b/ that $\lambda_0 \to \lambda_c \approx \frac{1}{\pi} \log \tau > 0$. Since the zero-energy value of the coupling is $\frac{\lambda}{\log \tau}$, the condition for the existence of a nontrivial L^2 solution is accordingly that this value be greater than $\frac{1}{\pi}$ (approximately).

For the nonlinear equation (11), one can apply the following theorem in bifurcation theory /Pimbley, 1969; Atkinson, 1987/:

> Let $m(x) = \lambda T(m; x)$, where T is a nonlinear, odd, thrice Fréchet-differentiable operator. Suppose that $T'(0; x)$ is compact and that λ_c^{-1} belongs to its (point) spectrum. Then λ_c is a bifurcation point of the nonlinear equation, such that the trivial solution splits into nontrivial ones as λ sweeps past λ_c.

In our case, the kernel corresponding to the Fréchet derivative, $T'(0, x)$, is $\{\max(x, y) \log[\tau + \max(x, y)]\}^{-1}$, and to assure compactness in the I.R., a cut-off must be used. This can take the form of a sharp truncation of the integration domain, or, more elegantly, of the insertion of a factor $y/(y+m_0^2)$, which transforms the Fréchet derivative of (11) into our quasilinear form (12).

4 Confinement

The hypothesized confinement property of QCD can be realized if the effective gluon propagator, $D_{\mu\nu}(k)$, behaves like k^{-4} as $k \to 0$. Such a behaviour has been made plausible by a nonperturbative (albeit cavalier) approximate treatment of the gluon propagator equation /Atkinson, 1984; Brown, 1989/. In 4-dimensional QCD, the regularization of this involves replacing k^{-4} by $\delta^4(k)$ /Munczek, 1986/, and this results in a new algebraic term in the equation for the quark mass-function. In place of the left-hand side of (11), we now have

$$m(x)\left\{1-\frac{\eta^2}{x+m^2(x)}\right\}, \tag{14}$$

where η^2 is a measure of the strength of the confining potential. The Fréchet derivative now fails to be compact if the term in parentheses has a zero in $[0,\infty)$. This certainly happens if $m(0) < \eta$, for the linearized integral equation is then a Fredholm equation of the third kind /Bart, 1973/. Indeed, the quasilinear variant of the equation now reads

$$m_0(x)\frac{x+m_0^2-\eta^2}{x+m_0^2} = \lambda\int_0^\infty \frac{dy}{\max(x,y)\log[\tau+\max(x,y)]}\frac{y m_0(y)}{y+m_0^2}; \tag{15}$$

and this is a homogeneous Fredholm equation of the second kind if $m_0 > \eta$, with a smallest eigenvalue, $\lambda = \lambda_c > 0$. When $m_0 < \eta$, there is a zero on the left-hand side; and from the work of Bart and Warnock, we know that there is a unique solution of the form

$$m_0(x) = \delta(x+m_0^2-\eta^2)+\tilde{m}_0(x), \tag{16}$$

where $\tilde{m}_0(x)$ is square integrable. A necessary condition that a square-integrable solution exist is that $m_0 > \eta$.

References

Atkinson, D. et al. /1984/ Jour. Math. Phys **25**, 2095 (Sec. 4)

Atkinson, D. /1987/ Jour. Math. Phys **28**, 2494 (Sec. 3)

Atkinson, D., Johnson, P.W. /1988a/ Phys. Rev. **D37**, 2290 (Sec. 2)

Atkinson, D., Johnson, P.W./1988b/ Phys. Rev. **D37**, 2296 (Sec. 3)

Bart, G.R., Warnock, R.L. /1973/ SIAM J. Math. Anal. **4**, 609 (Sec. 4)

Brown, N., Pennington, M.R. /1989/ Phys. Rev. **D39**, 2723 (Sec. 4)

Miransky, V.A. /1985/ Phys. Lett. **165B**, 401 (Sec. 3)

Munczek, H.J. /1986/ Phys. Lett. **B175B**, 215 (Sec. 4)

Pimbley, G.H. /1969/ Lecture Notes in Mathematics **104** (Springer, Berlin) (Sec. 3)

On the Scaling Behavior of the O(N) σ-Model

B. Bonnier and M. Hontebeyrie

Laboratoire de Physique Théorique, Unité Associée au CNRS UA 764,
Rue du Solarium, F-33170 Gradignan, France

Abstract

The magnetic susceptibility and the correlation length of the O(N) non-linear two-dimensional σ-model are constructed from their high temperature expansions at 14th order and asymptotic scaling. For N = 3 and N = 4, this determination is compared with Monte-Carlo data and used to predict the onset of scaling at finite coupling.

This work deals with some aspects of the two-dimensional non linear σ-model with O(N)-symmetry and action

$$S(\beta,N) = \frac{1}{2g}\int dx(\partial\sigma)^2 \qquad , \ \beta = \frac{2\pi}{Ng} \tag{1}$$

The interesting feature of this model for $N \geq 3$ is its property of asymptotic freedom, which constraints the large β behavior of the observables, like the magnetic susceptibility χ_N and the correlation length ξ_N,

$$\chi_N \approx C_N \beta^{-b} \exp(2a\beta) \quad , \quad a = N/(N-2) \tag{2a}$$

$$\xi_N \approx D_N \beta^{a-b} \exp(a\beta) \quad , \quad b = (N+1)/(N-2) \tag{2b}$$

through integration of the renormalization group (RG) equations (which leaves the constants C_N and D_N unspecified).

Because of this property of asymptotic scaling, also found in four-dimensional non-abelian gauge theories, the σ-model is a popular testing ground for ideas and methods, like lattice simulations where one tries to check the behavior (2-a,b) at large but finite β.

Unfortunately, in spite of the large size of the lattice involved - up to 400x400 - this goal is not reached yet, and the Monte-Carlo (MC) analyses are rather conflicting, especially for the O(3) model. In some simulations /Fox et al., 1982 ; Fukugita, Oyanagi, 1983 ; Shenker, Tobochnick, 1980 ; Shigemitsu, Kogut, 1981/ the onset of the asymptotic scaling is observed, in agreement with the RG exponential slopes of Eqs.(2a,b), or with a maller slope in other ones /Berg, Luscher, 1981 ; Koibuchi, 1988/ ($a \approx 2.5$ instead of 3 - see also /Drouffe, Flyvbjerg, 1988/), and finally others /Wolff, 1989/ do not see any scaling. The agreement is better in the O(4) case (see however /Seiler et al., 1988/) where RG scaling is observed /Shigemitsu, Kogut, 1981 ; Fukugita, Oyanagi, 1983 ; Heller, 1988/ in some range $2.4 \leq \beta \leq 3.5$. When the data allow for an extrapolation of the constants C_N and D_N of Eqs.(2a,b), one finds values in perfect agreement for the C_N but with some incoherence for the D_N as we recall below :

$$10^4 C_3 = \begin{cases} 1074 \pm 150 & /\text{Shenker, 1980}/ \\ 1072 \pm \ \ 75 & /\text{Fukugita, 1983}/ \end{cases} \qquad 10^4 D_3 = \begin{cases} 36 \pm 3 & /\text{Shigemitsu, 1981}/ \\ 28 \pm 1 & /\text{Fox, 1982}/ \end{cases}$$

$$10^4 C_4 = \begin{cases} 1250 \pm 120 & /\text{Fukugita, 1983}/ \\ 1280 \pm \ \ 20 & /\text{Heller, 1988}/ \end{cases} \qquad 10^4 D_4 = \begin{cases} 10 \pm 2 & /\text{Fukugita, 1983}/ \\ 6.7 \pm 0.2 & /\text{Heller, 1988}/ \end{cases}$$

If some of these previous incoherences may reflect various estimates of the finite size effects, one cannot exclude other effects, really present in the thermodynamic limit, which complicate the behavior (2a,b). It thus appears useful, in the analysis of MC data with respect to scaling and finite size effects, to have an independant determination of χ_N and ξ_N on the whole range of couplings.

In this work, we give a determination of χ_N and ξ_N provided by a resummation of their high temperature (HT) expansion which fulfills the asymptotic RG scaling of Eqs.(2a,b). As explained elsewhere /Butera, Comi,Marchesini, 1988 ; Bonnier, Hontebeyrie, 1989/, such a resummation is not obvious on the range $\beta \geq 2.5$, because of complex singularities $\beta_s(N)$

which have a small imaginary part for N=3, and reach as $N=\infty$ four symmetric values

$$\beta_s(N=\infty) = \pm\, \beta_\infty \exp(\pm i\pi/4) \quad , \quad \beta_\infty \approx 2.8578 \tag{3}$$

To overcome this difficulty, the most efficient method we have found among standard tools is a Padé resummation in a mapped variable which rejects the singularities $\beta_s(N)$ far from the physical region. This is achieved in the following way.

First, the β-plane with four cuts $\beta(\lambda)=\pm\lambda\beta_c \exp(\pm i\theta_c)$, $\lambda\geq 1$, is mapped onto the u-plane, cut for $u\leq -1$, through the mapping

$$\beta = \beta_c 2^{(3/\alpha-1)} \alpha^{-1/2} (\alpha-2)^{(\alpha-2)/2\alpha} u(1+u)^{(2-\alpha)/2\alpha} (2+u)^{-2/\alpha} \tag{4}$$

where $\theta_c=\pi(\alpha-2)/2\alpha$, $\alpha>2$. The branch points $\beta(\lambda=1)$ are thus identified with the effective singularities of χ_N (or ξ_N) which are mapped into u=-1, the physical region $\beta>0$ corresponding to u>0. We then consider the functions

$$\overline{\chi}_N(\beta) = \beta\, e^{-c\beta} \chi_N^d(\beta) \qquad c=2N/(N+1) \qquad d=(N-2)/N+1 \tag{5a}$$

$$\overline{\xi}_N(\beta) = \beta\, e^{-c'\beta} \xi_N^{d'}(\beta) \qquad c'=N \qquad d'=N-2 \tag{5b}$$

which from (2a,b) are just constant at infinity (we impose the RG asymptotic scaling). Inserting in Def.(5a,b) the 14th order HT expansions of χ_N and ξ_N given by /Luscher, Weisz, 1988/, we obtain the 14th order expansions of $\overline{\chi}_N$ and $\overline{\xi}_N$ in the <u>variable u</u>. This allows us to construct their [P,Q] Padé approximants in this variable, up to P+Q=14, and we focus on the diagonal ones which embody RG scaling.

We have checked /Bonnier, Hontebeyrie, 1989/ the reliability of this method in the soluble example $N=\infty$, taking the mapping parameters β_c and α at their exactly known value $\beta_c=\beta_\infty$, $\alpha=4$. In the O(3) and O(4) models they are unknown and we consider them as free parameters, staying in some vicinity of their asymptotic values. We fix them by requiring two conditions. The first one is to obtain a good stability of the Padé table : we then obtain various solutions with a remarkable convergence, the three last diagonal Padé (5,5), (6,6) and (7,7) being degenerate within 3/1000 at any coupling. All these solutions, corresponding to slightly different mapping parameters, fit the data below $\beta \leq 2.5$, and their dispersion begin to be sizable for $\beta \geq 2.5$. In order to select further among them, we thus impose as a second condition to fit the MC data around $\beta \approx 3$. We then obtain the following results. In the O(4) case, the scaling onset appears at $\beta \approx 3.5$ and

$$1240 < 10^4 C_4 < 1280 \,, \qquad 6.1 < 10^4 D_4 < 6.5 \qquad (6)$$

These values correspond to the choices $2.32 \leq \alpha \leq 2.47$ and $3.02 \leq \beta_c \leq 3.11$ and are found in agreement with /Heller, 1988/. In the O(3) case, the scaling onset appears only at $\beta > 3.7$, with the values

$$944 < 10^4 C_3 < 1065 \qquad 25 < 10^4 D_3 < 28 \qquad (7)$$

corresponding to $2.18 < 10^4 C_3 \leq 2.40$, $3.02 \leq \beta_c \leq 3.25$. This clearly favors /Fox et al., 1982/ against /Shigemitsu, Kogut, 1981/.

In addition, it appears possible to perform an exponential fit of the observables at lower coupling ($3. \leq \beta \leq 3.5$), with however effective slopes smaller than the asymptotic RG ones : this explains the results of /Berg, Luscher, 1981 ; Koibuchi, 1988/. On the other hand, at larger coupling, the MC data /Bender, Wetzel, Berg, 1986 ; Wolff, 1988/ are in disagreement with our representations and are thus suspected to be not properly corrected of finite size effects.

To summarize our results, we have shown that for the σ-model with the O(N)-symmetric standard action, there is some competition for $N \geq 3$ between the RG asymptotic behavior and the occurence of complex singularities, reminiscent of real phase transitions when $N \leq 2$. This fact complicates the evaluation of the RG slopes through MC measurements : on the range of couplings presently available, finite size effects remain important, and there is no reason to insist on possible violations of asymptotic freedom.

REFERENCES

Bender I., W. Wetzel, B. Berg /1986/ Nucl.Phys.B269, 389.

Berg B., M. Luscher /1981/ Nucl.Phys.B190 [FS3], 412.

Bonnier B., M. Hontebeyrie /1989/ Phys.Lett.B226, 361.

Butera P., M. Comi, G. Marchesini /1988/ Nucl.Phys.B300 [FS22], 1.

Drouffe J.M., H. Flyvbjerg /1988/ Phys.Lett. B206, 285.

Fox G., R. Gupta, O. Martin, S. Otto /1982/ Nucl.Phys.B205 [FS5], 188.

Fukugita M., Y. Oyanagi /1983/ Phys.Lett.123B, 71.

Heller U. /1988/ Phys.Rev.D38, 3834.

Koibuchi H. /1988/ Z.Phys.C39, 443.

Luscher M., P. Weisz /1988/ Nucl.Phys.B300 [FS22], 325.

Seiler E., I.O. Stamatescu, A. Patrascioiu, V. Linke /1988/ Nucl.Phys.B305 [FS23], 623.

Shenker S.H., J. Tobochnik /1980/ Phys.Rev.B22, 4462.

Shigemitsu J., J.B. Kogut /1981/ Nucl.Phys.B190 [FS3], 365.

Wolff U. /1989/ Phys.Lett.222B, 473.

A Conjecture on the Possible Exact (1+1)-Dimensional Solutions to the Discrete Boltzmann Models I

H. Cornille

Service de Physique Théorique de Saclay,
Laboratoire de l'Institut de Recherche Fondamentale
du Commissariat à l'Energie Atomique,
F-91191 Gif-sur-Yvette Cedex, France

ABSTRACT

The exact multidimensional exponential type solutions to the discrete Boltzmann models, which have been found, are sums of unidimensional similarity waves. Our goal is to obtain, at a linear level, a physical signature of the class of possible exact rational solutions. In the discrete kinetic theory, contrary to the continuous theory, the macroscopic conservation laws are included into the set of nonlinear equations for the densities. These conservation laws, being linear differential relations, underline the above recalled result. For a class of (1+1)-dimensional rational solutions with two exponential variables, satisfying at least two conservation laws, we prove that the possible solutions are sums of similarity waves. We first consider the class of models with three independent densities and two conservation laws. Second we investigate what happens when the two conservation laws of mass and momentum exist. Third we sketch briefly results for five densities and three conservation laws. Finally we conjecture that all exact, bounded (1+1)-dimensional solutions are sums of similarity waves.

I. INTRODUCTION

The completely integrable equations can be characterized either by their infinite number of conservation laws or by their associated linear systems. Is it possible that for some classes of nonintegrable equations, with a finite number of conservation laws, the determination of the possible exact solutions can be found at a linear level? The answer is yes for the nonlinear discrete kinetic equations and the linear level is provided by the physical conservation laws of mass, momentum and energy.

Discrete kinetic theory, pioneered twenty-five years ago by Broadwell[1] is presently a popular field of research[2]. The velocity $\vec{v}$ can only take discrete values $\vec{v}_i$, $|\vec{v}_i| = 1$, $i = 1, ..., 2p$ and many models have been constructed. To each $\vec{v}_i$ is associated a density N_i and the simplest exponential type solutions are the similarity shock waves

$$N_i = n_{oi} + n_i/D, \quad D = 1 + u, \quad u = d\,\exp(\rho t + \gamma x) \tag{1.1}$$

Multidimensional exact solutions, which have been recently found[3], are sums of similarity waves

$$N_i = n_{oi} + \sum_{j=1}^{J} n_{ji}/D_j, \quad D_j = 1 + u_j, \quad u_j = d_j \exp(\rho_j t + \gamma_j x) \tag{1.2}$$

with $J = 2,3$ for the (1+1) and (2+1)-dimensional solutions. Why are the exact solutions found for the nonlinear discrete models, like in linear theories, only sums of similarity solutions?

In the continuous kinetic theory, starting from the one particle distribution function $N(\vec{x}, \vec{v}, t)$ (space $\vec{x}$, time t) the linear conservation laws for the mass $\mathcal{M}(\vec{x}, t)$, the average velocity and the energy are obtained through integration over the velocity space. This means that the nonlinear microscopic equations, provided by the Boltzmann equation, do not include directly the macroscopic conservation laws relations. On the contrary, for the discrete kinetic theory, because of the discretization, microscopic nonlinear equations and macroscopic linear conservation laws are mixed together. The linear conservation laws (at least two must be present), acting like a filter, select the superposition of similarity waves as the only possible exact rational solutions. The aim of this paper is to provide some rigorous basis to this fact. We restrict our study to (1+1)-dimensional models.

Let us consider rational densities solutions with two independent exponential variables of the (1.2) type, $\rho_i\gamma_j \neq \rho_j\gamma_i$. We disregard the models with only one conservation law[4] and which violate momentum conservation law.

So we always assume two conservation laws and for instance consider the most popular models with three independent densities

$$N_{1t} + N_{1x} = N_{2t} - N_{2t} = -aN_{3t} = b(N_3^2 - N_1 N_2) \tag{1.3}$$

($a = b = 1$ for the $4\vec{v}_i$ model and 2 for the Broadwell one). Among the three equations, only one is nonlinear while the two linear ones are equivalent to the two conservation laws of mass and momentum. We recall the previous (1+1)-dimensional results coming from the existence of two linear relations[3,5] :

(i) Assuming denominators of the type $D = 1 + u_1 + u_2 + du_1u_2$, then necessarily $d = 1$ and the solutions are sums of similarity waves, (ii) For more general denominators D, including terms of the fourth order, a partial study indicates that only sums of similarity waves are possible, (iii) Adding the nonlinear term in (1.3), then only sums of two similarity waves survive.

In section 2 we study the possible (1+1)-dimensional rational solutions common to two independent linear relations

$$-\mathcal{V}_t - \mathcal{V}_x = \alpha_1 \mathcal{Z}_t + \beta_1 \mathcal{Z}_x = \alpha_2 \mathcal{W}_t + \beta_2 \mathcal{W}_x, \quad \alpha_i \neq \beta_i, \quad \alpha_i\beta_j \neq \alpha_j\beta_i, \tag{1.4}$$

with two independent exponential variables u_1, u_2 (1.2), a common denominator D

$$D = D_4 = 1 + \Sigma u_i \left(1 + d_{3i} u_i u_j + d_{4i} u_i^2 u_j\right) + d u_1 u_2 + d_4 u_1^2 u_2^2 \qquad (1.5)$$

and well-defined assumptions (called 1), 2), 3)). Mainly we assume that the solutions which represent probabilities are bounded and are not the trivial constants. Further the similarity waves are $(1 + u_j)^{-1}$ and when $u_i = 0$, $u_j \neq 0$ then the solutions reduce to the similarity one associated to u_j. If only sums of similarity waves can exist this means that D in (1.4-5) must be factorized:

$$D = (1 + u_1)(1 + u_2)(1 + d_4 u_1 u_2) \qquad (1.6)$$

In section 2 and Appendix B we prove this result. Further we verify in Appendix A that if the linear (1.4) relations are associated to a nonlinearity of the (1.3) type, then $(1 + u_j)^{-1}$ are the only bounded exponential similarity waves.

In section 3, as a particular case of the (1.4) relations we study the (1+1)-dimensional mass and momentum conservation laws

$$\mathcal{M}_t + \mathcal{V}_x = 0, \quad \mathcal{V}_t + \mathcal{W}_x = 0 \qquad (1.7)$$

With the same assumptions as in section 2, then all section 2 results for the linear (1.4) relations can be applied, and the solutions are still sums of similarity waves.

II.3. DENSITIES AND 2 INDEPENDENT LINEAR RELATIONS (TABLES 1-2-3)

In this section we study $\mathcal{V}$, $\mathcal{Z}$, $\mathcal{W}$ solutions of

$$-(\mathcal{V}_t + \mathcal{V}_x) = \alpha_1 \mathcal{Z}_t + \beta_1 \mathcal{Z}_x = \alpha_2 \mathcal{W}_t + \beta_2 \mathcal{W}_x \qquad (2.1a)$$

$$\alpha_i \neq \beta_i, \quad \alpha_i \beta_j \neq \alpha_j \beta_i \qquad (2.1b)$$

that are rational functions with two independent exponential variables.

$$\mathcal{V} = v_{00} + V/D, \quad \mathcal{Z} = z_{00} + Z/D, \quad \mathcal{W} = w_{00} + W/D$$

$$u_i = d_i \exp\left(\rho_i t + \gamma_i x\right), \quad \rho_i \gamma_j \neq \rho_j \gamma_i \qquad (2.2)$$

with $V(u_1, u_2)$, $Z(u_1, u_2)$, $W(u_1, u_2)$, $D(u_1, u_2)$ polynomials in u_1, u_2. We assume

1) $\mathcal{V}$, $\mathcal{Z}$, $\mathcal{W}$ are bounded in the u_1, u_2 plane (the degrees of V, W, Z are at most equal to the D one) and the denominator D is common.

2) The exponential similarity waves bounded on the x-axis are of the type $(1 + u_j)^{-1}$, if $u_i = 0$, $u_j \neq 0$ then the (1+1)-dimensional rational solutions reduce to

the u_j similarity waves

$$\mathcal{V}(u_i = 0, u_j) = \text{const}(1+u_j)^{-1}, \ \mathcal{Z} \longrightarrow \text{const}(1+u_j)^{-1}, \ \mathcal{W} \longrightarrow \text{const}(1+u_j)^{-1} \tag{2.3}$$

up to constants. We exclude $\mathcal{V}(u_i = 0, u_j) = \text{const}$, $\mathcal{Z} = \text{const}$, $\mathcal{W} = \text{const}$. A fortiori we exclude the trivial constant solutions: $\mathcal{V} = \text{const}$ or $\mathcal{Z} = \text{const}$ or $\mathcal{W} = \text{const}$.

3) D as in (1.5), and V, Z of the fourth order

$$\begin{aligned} V &= v_0 + \Sigma u_i(v_{1i} + v_{3i}u_iu_j + v_{4i}u_i^2u_j) + vu_1u_2 + v_4u_1^2u_i^2 \\ Z &= z_0 + \Sigma u_i(z_{1i} + z_{3i}u_iu_j + z_{4i}u_i^2u_j) + zu_1u_2 + z_4u_1^2u_i^2 \end{aligned} \tag{2.4}$$

and for W a similar polynomial with w_0, w_{1i}, w_{3i}, w_{4i}, w, w_4 parameters. We notice that in D terms like $u_i^m, m > 1$ are omitted. Otherwise when $u_j = 0$ D is reduced to $1 + u_i$ plus u_i^m terms not allowed by our assumption 2.

What could be the common solutions to (2.1a-b)? Calling $\xi = \rho t + \gamma x$, a similarity variable, and defining $p(k) = \alpha_k\rho + \beta_k\gamma, q = \rho + \gamma$. We see from $q\mathcal{V}_\xi + p(1)\mathcal{Z}_\xi = q\mathcal{V}_\xi + p(2)\mathcal{W}_\xi = 0$ that $V(\xi)$, $\mathcal{Z}(\xi) = \text{const} - q\mathcal{V}/p(1)$, $\mathcal{W}(\xi) = \text{const} - q\mathcal{V}/p(2)$ are common solution. For two independent $(1+u_j)^{-1}$ similarity solutions $\gamma_1\rho_2 \neq \gamma_2\rho_1$ we define

$$p_j(k) = \alpha_k\rho_j + \beta_k\gamma_j \quad k = 1,2, \quad q_j = \rho_j + \gamma_j, \quad p_i(k)q_j \neq p_j(k)q_i, \tag{2.5}$$

then any linear combination (omiting the trivial constants) of the type

$$\mathcal{V} = \sum_1^2 a_j(1+u_j)^{-1}, \ \mathcal{Z} = -\Sigma a_jq_j/p_j(1)(1+u_j), \ \mathcal{W} = -\Sigma a_jq_j/p_j(2)(1+u_j) \tag{2.6}$$

will be solutions. For rational solutions in u_1, u_2, in supplement to the former u_j we can have $u_3 = u_1u_2$ with $p_3 = p_1 + p_2$, $q_3 = q_1 + q_2, \ldots$ $u_{m+n} = u_1^mu_2^n$ with $p_{m+n}(k) = mp_1(k) + np_2(k)$, $q_{m+n} = mq_1 + nq_2$. For D of the fourth order and of the type (2,4) we can have sums of three similarity waves with $u_3 = u_1u_2$ while for higher degree of the denominator other $u_{m+n} = u_1^mu_2^n$ can give new components. If it is obvious that such superposition of similarity waves are solutions, *it is not obvious that they are the only ones.*

An interesting property occurs because only derivatives appear in the differential relations (2.1a-b). The constants v_{00}, w_{00}, z_{00} being arbitrary we can choose, two kinds of representations.

(I) Substracting const.D to the numerators we *can always have a monomial term present in D but missing in V, W, Z.* For instance if $d_4u_1^2u_2^2$ is present in D we can put $v_4 = z_4 = w_4 = 0$ and so on.

(II) Let us start in (2.4) with $\delta = v_0 - v_{11} - v_{12} \neq 0$ and perform the transformation $\mathcal{V} \longrightarrow \mathcal{V}' = \mathcal{V} + \delta D$. Then the new parameters $v_0' = v_0 + \delta$, $v_{1i}' = v_{1i} + \delta$ satisfy $v_0' - v_{11}' - v_{12}' = 0$. Consequently *we can always choose* $V = \Sigma v_{1i}(1 + u_i) + u_1 u_2 ...$, $Z = \Sigma z_{1i}(1 + u_i) + ...$, $W = \Sigma w_{1i}(1 + u_i) + ...$.

Substituting (2.2) into (2.1) gives a first relation

$$Z(\alpha_1 D_t + \beta_1 D_x) + V(D_t + D_x) - D(\alpha_1 Z_t + \beta_1 Z_x + V_t + V_x) = 0 \qquad (2.7a)$$

and a similar one (2.7b) with the changes $\alpha_1 \longrightarrow \alpha_2$, $\beta_1 \longrightarrow \beta_2$ and $Z \longrightarrow W$. (2.71-72) are polynomials in u_1, u_2 and each coefficient of a monomial term $u_i^m u_j^n$ (written (m,n) in the Tables) must be zero. Our method is simple, V being common to both relations, then we eliminate the W, Z parameters, leaving relations between the V parameters alone. Then we check whether these relations are compatible with the two sets of α_i, β_i values. In fact we seek the conditions on the V parameters such that V be a common solution. Other constraints occur for the (Z, V) parameters alone (or $w, w_{1i}, ...; z, z_{1i}, ...$) coming from the fact that the two exponential are independent: $p_i q_j \neq p_j q_i$ or $(\alpha_k - \beta_k)(\gamma_i \rho_j - \gamma_j \rho_i) \neq 0$. We obtain two fundamental properties.

Lemma 1 : If $A[mp_i + nq_j] + B[mq_i + nq_j] = 0$, $i \neq j$, $i = 1,2$, $m \neq n$, then necessarily $A = B = 0$. We note that the determinant is $(m^2 - n^2)(p_1 q_2 - p_2 q_1) \neq 0$.

Lemma 2 : If $A[m_1 p_i(k) + n_1 p_j(k)] + B[m_2 p_i(k) + n_2 p_j(k)] = 0$, $k = 1,2$, i and j fixed, $p_i(k)$ and q_i defined in (2.5), $m_1 n_2 \neq n_1 m_2$, then necessarily $A = B = 0$. We note that the determinant is $(m_1 n_2 - m_2 n_1)(\alpha_1 \beta_i - \alpha_2 \beta_2)(\rho_i \gamma_j - \rho_j \gamma_i) \neq 0$. This lemma is useful when A and B depend on the parameters of V.

A: Bounded exponential similarity waves deduced from nonlinear equations

In order to justify our choice of the exponential similarity waves $(1 + u)^{-1}$, $u = d \exp \xi$, $\xi = \rho t + \gamma x$ (a result which cannot be obtained from the linear relations alone), let us add the nonlinear part of the equations for the simplest (1+1) dimensional models. They have three or four independent densities satisfying

$$\begin{gathered} \mathcal{V}_t + \mathcal{V}_x = -\alpha_1 \mathcal{Z}_t - \beta_1 \mathcal{Z}_x = -\alpha_2 \mathcal{W}_t - \beta_2 \mathcal{W}_x = -\alpha_3 \mathcal{Y}_t - \beta_3 \mathcal{Y}_x = \mathcal{Y}\mathcal{Z} - \mathcal{V}\mathcal{W} \\ \alpha_i \neq \alpha_j, \quad \beta_i \neq \beta_j, \quad \alpha_i \beta_j \neq \alpha_j \beta_i \end{gathered} \qquad (2.8)$$

(three if $\mathcal{Y} = \mathcal{Z}$, $\alpha_3 = \alpha_1$, $\beta_3 = \beta_1$). *The study is done in Appendix A.* In the ξ variable, the derivatives of the densities are proportional. Let us call $D(\xi)$ the inverse of the densities, then $D(\xi)$ satisfies a Ricatti equation and the exponential type solutions are those chosen in our assumptions 1, 2, 3.

If the solutions are sums of similarity waves $\mathcal{V} = v_0 + \Sigma v_j/(1+u_j)$ $u_j = d_j\exp(\rho_j t + \gamma_j x)$, what is the maximum number of components compatible with the nonlinearity? In (1+1)-dimensions the physically relevant models with $|v_i| = 1$ *cannot have more than two independent conservation laws.* So we assume $\mathcal{Y} = \mathcal{Z}$ and only three densities. We cannot have sums with more than two similarity waves. *In the sequel we consider (2.1a-b) and never the nonlinear part of (2.8).*

B: D at most of the third order $D_3 = 1 + \Sigma u_i(1 + u_1u_2d_{3i}) + du_1u_2$.

In Table 1 the (m,n) relations for (2.7a) are written down with, for simplicity, p_i for $p_i(1)$. For (2.7b) we exchange the Z, W parameters and p_i means $p_i(2)$. We have defined

$$A_i = q_ip_j - q_jp_i \neq 0, \quad \bar{A} = \sum_i v_{1i}A_i/p_j \;\; j \neq i, \quad \bar{p} = p_1 + p_2, \quad \bar{q} = q_1 + q_2 \quad (2.9)$$

and for simplicity we write $\hat{z}_{1i}p_i$ for $z_{1i}p_i(1) + q_iv_{1i}$ and so on. *We choose representation* II: $v_0 = \Sigma v_{1i}$, $z_0 = \Sigma z_{1i}$.Three cases occur: (i) $d \neq 0$, $d_{3i} = 0$ or D of the second order, (ii) $d_{3i} \neq 0$, (iii) $d_{31} \neq 0$, $d_{32} = 0$ (without assumptions on d in the two last cases). The u_i or (1,0) relations $\hat{z}_0p_i = \hat{z}_{1i}p_i$ become for representation II $\hat{z}_{1i}p_j = 0$ leading to $z_{1i}p_i + v_{1i}q_i = v_{1i}A_j/p_j$.

Theorem 1 : For D of the second order ($d_{3i} = v_{3i} = z_{3i} = 0$), then only sums of two similarity waves are possible.

From (2,1) we find $p_j\hat{z} = p_jz + q_jv = 0$, lemma 1 giving $z = v = w = 0$, while (1,1) is $(d-1)\bar{A} = 0$. If $d \neq 1$ then $\bar{A} = 0$ or $p_1(k)v_{11} = p_2(k)v_{12}$, $k = 1,2$, being relations for V alone, we apply lemma 2 and find $v_{1i} = 0$ or $V = Z = W = 0$. For $d = 1$ we obtain

$$\mathcal{V} = \Sigma v_{1i}/(1+u_j), \; \mathcal{Z} = -\Sigma v_{1i}q_j/p_j(1)(1+u_j), \; \mathcal{W} = -\Sigma v_{1i}q_j/p_j(2)(1+u_j) \quad (2.9)$$

and the solutions are sums of two similarity waves.

Theorem 2 : *D of the third order* (not reducing to a second order) *does not lead to (1+1)-dimensional solutions*

(2,1) and (3,1) give $d_{31}d_{32}\bar{A} = 0$. If $d_{31}d_{32} \neq 0$ then $\bar{A} = 0$ or $v_{1i} = 0$, $i = 1,2$ and $V(u_i = 0, u_j \neq 0) = 0$ violates our assumption 2. If $d_{31} \neq 0$, $d_{32} = 0$, $\bar{A} \neq 0$, we can choose $v_{32} = z_{32} = w_{32} = 0$ (assumption 1) but $v_{31} \neq 0$, $Z_{31} \neq 0$, $W_{31} \neq 0$. We find $p_1\hat{z} = 0$ or $zp_1 = -q_1v$, $p_2\hat{z} = vA_2/p_1$ with (2,1) $i = 2$; $d = 1 + d_3$ with (2,1); $d_{31}\bar{A} = p_2\hat{z}$ or $d_{31}v_{11}p_1(k) = (v - d_{31}v_{12})p_2(k)$. With lemma 2 we get $v_{11} = 0$, $v = d_{31}v_{12}$ while (2,2), (3,1) gives $p_j\hat{z}_{31} = 0, j = 1,2$ or $v_{31} = z_{31} = w_{31} = 0$. Then when $u_1 = 0$, $V/D = v_{12}$ violates assumption 2. In fact the solution obtained is a similarity wave : $\mathcal{V} = v_{00} + v_{12}/(1+u_1)$ with

$$V = v_{12}(1 + u_2 + d_{31}u_1u_2),\ v_{12}D = V(1+u_1),\ Z = -qV/p(1), W = -qV/p(2) \tag{2.10}$$

What happens if we give up our assumptions 1 and 2 ? First in the $d_{31}d_{32} \neq 0$ case let us go on with the $v_{1i} = 0$ solution. After a trivial algebra we still find a similarity wave in the exponential variable u_1u_2

$$\mathcal{V} = vu_1u_2/(1+du_1u_2),\ \mathcal{Z} = -\mathcal{V}\bar{q}/\bar{p}(1),\ \mathcal{W} = -\mathcal{V}\bar{q}/\bar{p}(2) \tag{2.11}$$

where we have omitted the constants. Second in the above $d_{31} \neq 0$, $d_{32} = 0$ case let us assume $v_{32} \neq 0$, $z_{32} \neq 0$, $w_{32} \neq 0$. Then (2,1), (3,1) lead back to $v_{32} = .. = 0$. As a last case, let us assume D of the second order but V, Z, W of the third one $v \neq 0$, $v_{3i} \neq 0, \ldots$ After trivial algebra we find solutions $\mathcal{V} = vu_1u_2$, $\mathcal{Z} = -\bar{q}\mathcal{V}/\bar{p}(1)$, $\mathcal{W} = -\bar{q}\mathcal{V}/\bar{p}(2)$ which are not rational functions.

C: D of the fourth order $D = D_4 = D_3 + \Sigma d_{4i}u_i^3u_j + d_4u_1^2u_2^2$ (Tables 2-3)

If only sums of similarity waves can be solutions, then D must be factorized and is of the type

$$D_4 = (1+u_i)(1+u_i)(1+d_4u_1u_2) \tag{2.12}$$

This is the result that we prove in Appendix B.

We must distinguish between five different cases: i) $d_4 = 0$, $d_{4i} \neq 0$; ii) $d_4 = 0$, $d_{41} \neq 0$, $d_{42} = 0$; iii) $d_4 \neq 0$, $d_{41} \neq 0$, $d_{42} = 0$; iv) $d_4 \neq 0$, $d_{4i} \neq 0$; v) $d_4 \neq 0$, $d_{4i} = 0$. We expect that only the last case should lead to (1+1)-dimensional solutions with factorized D while the others lead either to similarity waves, to trivial constant solutions or to impossibilities. In each case, all other D parameters as well as V, W, Z parameters can be either zero or not. The discussion of all possible subcases is tedious and all details are given in Appendix B.

Here we sketch briefly the main results. For the *first four possibilities* we use *representation* I with $v_0 \neq \Sigma v_{1i}, \ldots$ at the starting point, allowing to eliminate some highest monomial term of the numerators. The coefficients of $u_i^m u_j^n$ in (2.7a) are written down like (m,n) in Table 2. Those of (2.7b) are deduced with the change $p_i(1) \longrightarrow p_i(2)$ and the parameters of W instead of those of Z. For the *last v) case*, the (m,n) relations are written down *in Table 3 with the representation II* at the starting point, $v_0 = \Sigma v_{1i}\ldots$ The results for the first four cases are provided by *Theorems 3,4,5,6 of Appendix B.* We find either constants, impossibilities or unidimensional similarity waves. We notice that these similarity waves have been found because we have allowed a violation of a part of our assumption 2). Namely $V(u_i = 0,\ u_j \neq 0) = \textit{constant}$ is verified for all these similarity waves. On the contrary the last case $d_4 \neq 0$, $d_{4i} = 0$. *Theorem 7* in *Appendix B*, is the only one leading to (1+1)-dimensional solutions and they are sums of two or three similarity waves. D is of the (2.12) type while V depends on three arbitrary parameters v_{1i}, v_4 leading to

$$\mathcal{V} = v_{00} + \sum_1^2 v_{1i}/(1+u_j) + v_4 u_1 u_2/(1+d_4 u_1 u_2),$$

and $\mathcal{Z}$, $\mathcal{W}$ of the same type are written down in (B.9).

III. MASS AND MOMENTUM CONSERVATION LAWS (TABLE 4)

In (1+1)-dimensions we define the total mass $\mathcal{M}$, the momentum $\mathcal{V}$ (for discrete models $\mathcal{M} = \Sigma N_i$, $\mathcal{V} = \Sigma N_i \vec{v}_i$) and write down two conservation laws

$$\mathcal{V}_x + \mathcal{M}_t = 0 \quad \mathcal{V} = v_{00} + V/D, \qquad \mathcal{M} = m_{00} + M/D \tag{3.1a}$$

$$\mathcal{V}_t + \mathcal{W}_x = 0 \quad \mathcal{W} = w_{00} + W/D \tag{3.1b}$$

We still assume that D, V, M, W are polynomials in two independent exponential variables $u_i = d_i \exp(\rho_i t + \gamma_i x)$. If we compare with the two differential relations (2.7a-b) of section 2 we find the correspondence

$$\begin{aligned}
\mathcal{Z} \longrightarrow \mathcal{M},\ & p_i(1) = \rho_i, q_i(1) = \gamma_i,\ \alpha_1 = 1,\ \beta_1 = 0, A_i(1) = \gamma_i \rho_j - \gamma_j \rho_i, \\
& \bar{A}(1) = \sum_i v_{1i} A_i(1)/\rho_j\ j \neq i;\ p_i(2) = \gamma_i,\ q_i(2) = \rho_i,\ \alpha_2 = 0, \beta_2 = 1, \\
& A_i(2) = A_j(1), \\
& \bar{A}(2) = \sum_i v_{1i} A_i(2)/\gamma_j\ ; \alpha_1\beta_2 = 1 \neq \alpha_2\beta_1 = 0
\end{aligned} \tag{3.2}$$

With the same above assumptions 1, 2, 3 for the possible (1+1)-dimensional relations we obtain the results of section 2: only similarity waves (violating assumption 2 or sums of two or three similarity waves are possible for denominators at most of the fourth order in u_1, u_2. The main difference with section 2 is that we do not have the nonlinear counterpart of (3.1a-b). Consequently we cannot justify our assumption that the similarity waves are of the type $(1+u)^{-1}$ with $u = \exp(\rho t + \gamma x)$.

In order to illustrate the results of section 2 to (3.1a-b) we consider D of the third order. Applying the correspondence (3.2), the reader can perform as well all the tedious calculations of Appendix B to the present case. In Table 4, *choosing* (like in Table 1) *the representation* II $m_0 = \Sigma m_{1i}, \ldots$ we write down the (s, n) coefficient relations of $u_i^s u_j^n$ coming from

$$V\ D_x + M\ D_t - D(V_t + M_t) = 0 \tag{3.3a}$$

$$W\ D_x + V\ D_t - D(W_x + V_t) = 0 \tag{3.3b}$$

For D of the second order $d_{3i} = v_{3i} = w_{3i} = m_{3i} = 0$, we find from (2,1) $\rho_j \hat{m} = 0$, $\gamma_j \hat{w} = 0$ or $m = w = v = 0$, while (1,1) gives $(d-1)\bar{A}(1) = (d-1)\bar{A}(2) = 0$. If $d = 1$ then the solutions are sums of two similarity waves. If $d \neq 1$ then $\bar{A}(1) = \bar{A}(2) = 0$ or $v_{1i}\rho_i = v_{1j}\rho_j$, $v_{1i}\gamma_i = v_{1j}\gamma_j$ which lead to $v_{1i} = 0$, $V = M = Z = 0$.

For D of the third order we still have $d_{31}d_{32}\bar{A}(k) = 0$, $k = 1,2$ leading to the same discussion as in section 2. If $d_{31}d_{32} \neq 0$ then $\bar{A}(k) = 0$, $v_{1i} = 0$, $V/D = 0$ for $u_i = 0$ which violates assumption 2. We consider the last case $d_{31} \neq 0$, $v_{31} \neq 0, \ldots$, $d_{32} = v_{32} = m_{32} = w_{32} = 0$; then (2,1) $i = 2$ gives $\rho_1 \hat{m} = \gamma_1 \hat{w} = 0$, $v = -\rho_1 m/\gamma_1 = -\gamma_1 w/\rho_1$, and either $d = 1 + d_{31}$ or $\bar{A}(j) = 0$, $j = 1,2$, $v_{1i} = 0$, $V/D = 0$ for $u_i = 0$ which violates assumption 2. Then (1,1) gives $d_{31}\bar{A}(1) = \rho_2 m + \gamma_2 v = vA_2(1)/\rho_1$, $d_{31}\bar{A}(2) = \gamma_2 w + \rho_2 v = -vA_2(1)/\gamma_1$ or $\rho_1 v_{11} d_{31} + \rho_2(v - d_{31}v_{12}) = \gamma_1 v_{11} d_{31} + \gamma_2(v - d_{31}v_{12}) = 0$. As in section 2 we deduce $v_{11} = 0$, $v = d_{31}v_{12}$. Then (2,2), (3,1) give $\rho_j \hat{m}_{31} = \gamma_j \hat{w}_{31} = 0$, $j = 1,2$ or $\rho_j m_{31} + \gamma_j v_{31} = \gamma_j w_{31} + \rho_j v_{31} = 0$ or $v_{31} = m_{31} = w_{31} = 0$. Consequently $V/D \longrightarrow v_{12}$ when $u_1 = 0$ and violates assumption 2. The solution is a similarity wave $V/D = v_{12}/(1 + u_1)$. Giving up assumptions 1 and 2 we find the same results as in section 2, for instance similarity waves of the (2.11) type in the exponential variable $u_1 u_2$. We can go on but the discussion is the same as in section 2 leading to the results that only sums of similarity waves can be (1+1)-dimensional solutions.

Another way to study (3.1a-b) is to look at the compatibility condition:

$$\mathcal{M}_{tt} = \mathcal{W}_{xx} \tag{3.4}$$

While sums of similarity waves $\mathcal{M} = \Sigma m_i/(1 + u_i)$, $\mathcal{W} = \Sigma m_i(\rho_i/\gamma_i)^2/(1 + u_i)$ are possible, we have verified that solutions with $D = 1 + \Sigma u_i + du_1u_2$, $d \neq 1$, and numerators $\mathcal{M}, \mathcal{W}$ of the first order in u_i, are incompatible with (3.4) (Appendix C). We notice that (3.4) is obtained by applying the operator ∂_t to (3.1a) and ∂_x to (3.1b). Consequently if we add an arbitrary x-dependent function to (3.1a) and an arbitrary t-dependent one to (3.1b) we still obtain (3.4). It is why, in the present paper, instead of looking at the possible compatibility relations which could introduce spurious solutions, we have thought preferable to work directly with the original equations.

IV. CONCLUSION

Our main result is associated with the physical linear differential part of discrete kinetic equations. If *three densities satisfy two independent linear conservation laws relations* and if we require the assumptions 1, 2, 3 of section 2, then the *only possible exact* (1+1)-dimensional rational solutions, with two exponential variables, are *sums of similarity waves*. We conjecture that this result is general and not restricted to fourth order denominators. For instance adding to (1.5) terms of the fifth order: $\Sigma d_{53i} u_i^3 u_j^2 + d_{54i} u_i^4 u_j$ we conjecture that D_5 must be $(1 + u_1)$ $(1 + u_2)$ $(1 + d_{53i} u_i^2 u_j)$ with i fixed. Further for D_6 calling $D_2 = (1 + u_1)\,(1 + u_2)$ then we find only $D_2(1 + u_1^2 u_2^2 d_6)$ and so on still assuming $D = 1 + \Sigma u_i + u_1 u_2$ $Q(u_1, u_2)$ with Q a polynomial as in section 2. Moreover, for D of the (1.5) type, giving up assumption 2 for the possible solutions we have verified that no other classes of (1+1)-dimensional solutions occur.

However other models with more densities and conservation laws exist. For instance the Cabannes[5] model has five densities and three linear relations, each of them linking two densities. We still find that the possible solutions are sums of similarity waves (Appendix D).

In conclusion these restrictions on the possible exact solutions come from the physical property that in discrete kinetic theory the linear conservation laws are included.

REFERENCES

(1) Broadwell J.E., Phys. Fluids **7**, 1243 (1964).

(2) For books and review articles on the discrete kinetic theory: Gatignol R., "Lecture Notes in Physics" **36** Springer, Berlin (1975); TTSP **16**, 809 (1987); Platkowski T., Illner R., SIAM Review **30**, 212 (1988); Cercignani C., Illner R., Shinbrot M., Comm. Math. Phys. **114**, 687 (1988), "Discrete Kinetic Theory, Lattice Gas Dynamics and Foundations of Hydrodynamics" Edit. Monaco R. World Scientific Publishing, Singapore (1989); d'Humières D., "Bibliography on Lattice Gases and related Topics", Les Houches School February (1989).

(3) For a recent review article: "Exact Solutions of the Boltzmann Equation", Cornille H., Les Houches School March (1989), JMP **30**, 789 (1989).

(4) For the $2\vec{v}_i$ models with only one conservation law (Platkowski T., Jour. Méc. Théo. Ap. **4**, 555 (1985)) I recall the known results for the exact solutions. There exists a completely soluble model, the Ruijgrook-Wu (Physica **13A**, 401 (1982)) model. For the other $2\vec{v}_i$ models (Cornille H., JMP **28**, 1567 (1987)) the solutions are still sums of similarity waves but for the proof we must include a part of the nonlinearity and so the possible class of solutions is not obtained at a linear level. The Ruijgrook-Wu model is the only one with other solutions than sums of similarity waves (result obtained with T.T.Wu in Appendix 2.3).

(5) Cornille H., J. Stat. Phys. **48**, 789 (1987); "Third International Workshop on Mathematical Aspects of Fluid and Plasma Dynamics", Lect. Notes Math., Ed. Toscani G., (1989).

(6) Cabannes H., J. Mech. **14**, 705 (1975); Cabannes H. and Tiem D.M., Complex Systems **1**, 574 (1987).

APPENDIX A - EXPONENTIAL TYPE SIMILARITY WAVES

1. Similarity waves : We assume that 3 or 4 densities ($\mathcal{Y} = \mathcal{Z}$ if $\alpha_3 = \alpha_1$, $\beta_3 = \beta_1$) satisfy

$$-\mathcal{V}_t - \mathcal{V}_x = \alpha_1 \mathcal{Z}_t + \beta_1 \mathcal{Z}_x = \alpha_2 \mathcal{W}_t + \beta_2 \mathcal{W}_x = \alpha_3 \mathcal{Y}_t + \beta_3 \mathcal{Y}_x = \mathcal{V}\mathcal{W} - \mathcal{Y}\mathcal{Z} \quad (A.1)$$

We seek similarity waves with the variable expξ, $\xi = \rho t + \gamma x$, bounded on the ξ-axis. From the linear relations we see that $\mathcal{V}_\xi$, $\mathcal{Z}_\xi$, $\mathcal{W}_\xi$, $\mathcal{Y}_\xi$ are proportional. We can write $\mathcal{V} = v_0 + v/D$, $\mathcal{Z} = z_0 + z/D$, $\mathcal{W} = w_0 + W/D$, $\mathcal{Y} = y_0 + Y/D$, $D = D(\xi)$ and get a Ricatti equation

$$aD_\xi + a_0 + a_1 D + a_2 D^2 = 0,$$
$$a_0 = yz - vw, \quad a_1 = yz_0 + zy_0 - vw_0 - wv_0, \quad a_2 = y_0 z_0 - v_0 w_0$$
$$a = v(\rho + \gamma) = -z(\alpha_1 \rho + \beta_1 \gamma) = -w(\alpha_2 \rho + \beta_2 \gamma) = -y(\alpha_3 \rho + \beta_3 \gamma)$$

(i) If $a_2 = 0$, the solution is a constant plus are exponential $D = -a_0/a_1 + d$ $\exp(-a_1 \xi / a)$ while the choice $a = -a_1 = a_0$ leads to $D = 1 + d$ expξ.

(ii) If $a_2 \neq 0$ putting $D = (a/a_2)\partial_\xi \log E$ then E is a sum of two exponentials $\exp(\lambda_i \xi)$, $i = 1, 2$. If $\lambda_1 \neq \lambda_2$, coming back to D^{-1} we find $D^{-1} = c_1 + c_2/(1 + \exp(\lambda_1 - \lambda_2)\xi)$, c_i =constants. If $\lambda_1 = \lambda_2 = -a_1/2a$ the two independent solutions are exp$\lambda\xi$, ξexp$\lambda\xi$ leading to power type solutions for D which are excluded.

2. Sums of similarity waves : The physically relevant (1+1)-dimensional models ($|\vec{v}_i| = 1$) having only two conservation laws we restrict our study to $\mathcal{Y} = \mathcal{Z}$. We assume

$$\mathcal{V} = v_0 + \Sigma v_j / D_j, \quad \mathcal{W} = w_0 + \Sigma w_j / D_j, \quad \mathcal{Z} = z_0 + \Sigma z_j / D_j,$$
$$D_j = 1 + d_j \exp(\rho_j t + \gamma_j x), \quad \alpha_i \neq \beta_i,$$

and $\alpha_1 \beta_2 \neq \alpha_2 \beta_1$. We substitute into (A.1), define $x_j = w_j / v_j$, $\bar{z}_j = z_j / v_j$ and obtain from the linear part : $\bar{z}_j = a x_j / (b + c x_j)$, $a = \alpha_1 - \beta_1$, $b = \alpha_2 - \beta_2$, $c = \alpha_1 \beta_2 - \alpha_2 \beta_1$, while the coefficient of $1/D_i D_j$ $i \neq j$ into the collision term gives $x_i + x_j = 2 \bar{z}_i \bar{z}_j$. For a superposition of three similarity waves we must have three different x_1, x_2, x_3 satisfying: $F(x_i, x_j) = b^2 (x_i + x_j) + c^2 (x_i x_j^2 + x_j x_i^2) + bc(x_i^2 + x_j^2) + 2 x_i x_j (bc - a)^2 = 0$ or

$$O = b^2 + c^2 x_i (x_i + x_j + x_k) + bc(x_j + x_k) + 2 x_i (bc - a^2) \quad (A.2)$$

We find $x_i + x_j + x_k = (2a^2 - bc)/c^2$ which substituted into (A.5) gives $abc = 0$ or an impossibility. In conclusion at most a sum of 2 similarity waves is possible.

APPENDIX B - 3 DENSITIES AND 2 CONSERVATION LAWS

We recall the relations for the ansatz rational solutions $\mathcal{V} = V/D$, $\mathcal{Z} = Z/D$, $\mathcal{W} = W/D$,

$$X(\alpha_k D_t + \beta_k D_x) + V(D_t + D_x) = D(\alpha_k X_t + \beta_k X_x + V_t + V_x),$$
$$\alpha_i \neq \beta_i,\ \alpha_i\beta_j \neq \alpha_j\beta_i,\ \ k = 1,2$$

$$D = 1 + \Sigma u_i + d u_1 u_2 + \Sigma u_i^2 u_j(d_{3i} + u_i d_{4i}) + d_4 u_1^2 u_2^2,$$
$$X = x_0 + \Sigma x_{1i} u_i + x u_1 u_2 + \Sigma u_i^2 u_j(x_{3i} + u_i x_{4i}) + x_4 u_1^2 u_2^2 \qquad (B.1)$$

$$k = 1 \quad X, x_0, \ldots \longrightarrow V, v_0, v_{1i}, v, v_{3i}, v_{4i}, v_4 \text{ and } Z, z_0, z_{1i}, z_1, z_{3i}, z_{4i} z_4$$
$$k = 2 \quad X, x_0, \ldots \longrightarrow V, v_0, v_{1i}, v, v_{3i}, v_{4i}, v_4 \text{ and } W, w_0, w_{1i}, w, w_{3i}, w_{4i}, w_4$$

with $u_i = \exp(\rho_i t + \gamma_i x)$, $i = 1,2$, $\rho_i\gamma_j \neq \rho_j\gamma_i$ in 1+1 dimensions. We define $p_i(k) = \alpha_k\rho_i + \beta_k\gamma_i$, $q_i = \rho_i + \gamma_i$ with $A_i = q_i p_j - q_j p_i \neq 0$, $i \neq j$ and two main properties

$$A(mp_i(k) + np_j(k)) + B(mq_i + nq_j) = 0$$
$$k \text{ fixed},\ i = 1,2\ i \neq j,\ m^2 \neq n^2 \longrightarrow A = B = 0 \qquad (B.2)$$

$$A(m_1p_1(k) + n_1p_2(k)) + B(m_2p_1(k) + n_2p_2(k)) = 0,$$
$$k = 1,2,\ m_1n_2 \neq m_2n_1 \longrightarrow A = B = 0 \qquad (B.3)$$

The Tables 2-3 are written with Z, V parameters and p_i means $p_i(k = 1)$ while the same relations exist for W, V parameters with p_i meaning $p_i(k = 2)$. We seek relations for V parameters alone valid with $p_i(k)$, $k = 1,2$ and apply (B.3). Tables 2-3 correspond respectively to representations I $v_0 \neq \Sigma v_{1i}$, $z_0 \neq \Sigma z_{1i}, \ldots$ and II $v_i = \Sigma v_i, z_i = \ldots$. For any z_α parameters we write $\hat{z}_\alpha f(p_i, p_j)$ for $z_\alpha f(p_i, p_j) + v_\alpha f(q_i, q_j)$ and write down important tools:

(i) If $\hat{z}_\alpha(m_1p_1 + n_1p_2) = 0$ then $\hat{z}_\alpha(m_2p_1 + n_2p_2) = v_\alpha(m_1n_2 - n_2m_1)\ A_1/(m_1p_1 + n_1p_2)$, $p_i = p_i(k = 1)$.

(ii) If $\hat{z}_\alpha(m_ip_1 + n_ip_2) = 0$, $i = 1,2$, if $m_1n_2 \neq m_2n_1$ then $v_\alpha = z_\alpha = 0$.

(iii) If $\hat{z}_\alpha(m_1p_1 + n_1p_2) = \hat{z}_\beta(s_1p_1 + t_1p_2) = 0$, if $\hat{z}_\alpha(m_2p_1 + n_2p_2) = \hat{z}_\beta(s_2p_1 + t_2p_2)$, if $m_1n_2 \neq m_2n_1$, $s_1t_2 \neq s_2t_1$, $s_1n_1 \neq t_1n_1$ then $v_\alpha = v_\beta = z_\alpha = z_\beta = w_\alpha = w_\beta = 0$.

B.1: Case $d_4 = v_4 = w_4 = z_4 = 0$, $d_{4i} \neq 0$, $i = 1,2$ and the choice $v_{42} = z_{42} = w_{42} = 0$, Table 2.

Theorem 3: The only possibility is $V = W = Z = 0$.
From (4,4), (5,2)$i = 1,2$, (4,3)$i = 1,2$, we get $(p_2 - p_1)\hat{z}_4 = \hat{z}_{32}p_2 = \hat{z}_{31}(2p_2 -$

$p_1) = 0$, $p_1 d_{41}\hat{z}_{41} = p_1 d_{31}\hat{z}_{31}$, $(2p_1 - p_2)(\hat{z}_{32}d_{41} - \hat{z}_{41}d_{32}) = 0 \longrightarrow (p_2 - p_1)2v_{31} + d_{31}v_{41}(2p_2 - p_1) = 0$, $(p_2 - p_1)\, d_{41}v_{32} + d_{32}v_{41}p_2 = 0 \longrightarrow v_{31} = v_{32} = d_{31}\, v_{41} = d_{32}v_{41} = 0$.
From $(4,2)i = 1,2$, $(4,1)i = 1$, $(3.2)i = 2$ we get $\hat{z}p_2 = 0$, $dp_1\hat{z}_{41} = d_{41}p_1\hat{z}$, $(2p_1 + p_2)(d_{41}\hat{z}_{11} - \hat{z}_{41}) = 0 \longrightarrow dv_{41}p_2 + d_{41}v(p_2 - p_1) = 0$, $(p_2 - p_1)2d_{41}v_{11} = 3v_{41}p_2 \longrightarrow v_{41} = v_{11} = v = 0$. Finally from $(4,1)i = 2$, $(3,2)i = 1$, $(1,0)$ we get $p_i\hat{z}_{12} = p_i\hat{z}_0 = 0$, $i = 1,2 \longrightarrow v_0 = v_{12} = 0 \longrightarrow V = 0$.

B.2: Case $d_4 = v_4 = \ldots = 0$, $d_{41} \neq 0$, $d_{42} = v_{42} = \ldots = 0$ and the choice $v_{41} = z_{41} = w_{41} = 0$.

Theorem 4: The only possible solutions are similarity waves (Table 2).
From $(4,1)i = 1$, $(3,2)i = 1$, $(1,0)i = 1,2$ we get $(2p_1 + p_2)\hat{z}_{11} = p_1\hat{z}_{12} = 0$, $p_2 z_2 + q_2 v_0 = -A_1 v_{12}/p_1$, $p_1 z_0 + q_1 v_0 = A_1 v_{11}/(2p_1 + p_2)$, $A_1 = q_1 p_2 - q_2 p_1 \longrightarrow (v_0 - v_{12})(2p_1 + p_2) = v_{11}p_2 \Longrightarrow v_{11} = 0$, $v_0 = v_{12}$ and $V/D = v_0$ when $u_1 = 0$ which violates assumption 2. However let us go on. From $(4,3)i = 1$, $(2,2)$ we get $p_j\hat{z}_{32} = 0$, $j = 1,2$ or $v_{32} = 0$; from $(4,2)i = 1$, $(3,2)i = 2$ we get $d_{32}\bar{p}\hat{z}_{12} = 0$ and $d_{32}v_{12} = 0$. From (3.3) we get $d_{32}p_2\hat{z}_{31} = 0$ and $d_{32}v_{31} = 0$. If $d_{32} \neq 0$, then $v_{11} = v_{12} = v_0 = v = v_{31} = v_{32} = V = 0$ and so necessarily $d_{32} = 0$. From (1,1) $(3,1)i = 1$, $(2,1)i = 1$ we get $(d-1)p_2\hat{z}_{12} = \hat{z}p_2$ and $v_{12}(d-1) = v$, $d_{41}p_2\hat{z}_{12} = \hat{z}_{31}\bar{p}$ and $d_{41}v_{12} = v_{31}$, $p_2(d_{31}\hat{z}_{12} - \hat{z}_{31} - \hat{z}) = 0$ or $d_{31}v_{12} = v_{31} + v$. Finally

$$\begin{gathered} V = v_0(1 + u_2 + d_{41}u_1^2 u_2 + (d-1)u_1 u_2),\ \mathcal{V} = V/D = v_0/(1 + u_1), \\ z = -q_1\mathcal{V}/p_1(1),\ \mathcal{W} = -q_1\mathcal{V}/p_1(2) \end{gathered} \tag{B.4}$$

is a similarity wave and not an (1+1)-dimensional solution.

B.3: Case $d_{42} = v_{42} = \ldots = 0$, $d_{41} \neq 0$, $d_4 \neq 0$, the choice $v_{41} = \ldots = 0$ but $v_4 \neq 0$, Table 2.

Theorem 5: The only possible solutions are either $\mathcal{V} = \ldots = 0$ or similarity waves.
From $(5,3)i = 1$, $(5,2)i = 1$, $(4,2)i = 1$, $(4,1)i = 1$ we get $(p_2 - p_1)\hat{z}_4 = p_1\hat{z}_{31} = p_1\hat{z} = (2p_1 + p_2)\,\hat{z}_{11} = 0$ while $(4,3)i = 1,2$ and (3,3) lead to $\hat{z}_{32}p_1 d_4 = \hat{z}_4 p_1 d_{32}$, $(2p_1 - p_2)\hat{z}_{32}d_{31} + p_2\hat{z}_{31}d_4 = \hat{z}_4 p_1 d_{41}$, $d_{31}(p_2 - p_1)\hat{z}_{32} - d_{32}p_2\hat{z}_{31} + \bar{p}d\hat{z}_4 = p_2 d_4\hat{z}$ or: $v_4 d/d_{31} = 0$, $v_4(d_{32}/d_4 - d_{31}/d_{41}) = 0$, $v_{32} - d_4 v_{31}/d_{41} + v_4 d_{32}/d_4 = 0$, $v_{31}(d_4/d_{41} - d_{32}/d_{31}) = vd_4/d_{31}$.

(i) If $v_4 = 0$ we find two similarity waves

$$\begin{gathered} V = v_0(1 + u_1 + u_2),\ \mathcal{V} = v_0/(1 + d_4 u_1^2 u_2),\ (2p_1 + p_2)\hat{z}_0 = 0, \\ \mathcal{Z} = -\mathcal{V}(2q_1 + q_2)/(2p_1(1) + p_2(1)),\ \mathcal{W} = -\mathcal{V}(2q_1 + q_2)/(2p_1(2) + p_2(2)) \end{gathered} \tag{B.5}$$

$$\begin{gathered} V = v_0(1 + u_2 + d_{41}u_1^2 u_2 + d_4 u_2^2 u_1),\ \mathcal{V} = v_0/(1 + u_1),\ p_1\hat{z}_0 = 0, \\ \mathcal{Z} = -\mathcal{V}q_1/p_1(1),\ \mathcal{W} = -\mathcal{V}q_1/p_2(1) \end{gathered} \tag{B.6}$$

(ii) If $v_4 \neq 0$ we have $d = v = 0$, $d_{32}d_{41} = d_{31}d_4$, and from (1,1), (2,1)$i = 1$, (3,2)$i = 1$ we get $(p_1 - p_2)\hat{z}_{12} = 3p_1\hat{z}_{11}$, $2p_1\hat{z}_{11} + p_2\hat{z}_{12} = \hat{z}_{31}p_2/d_{31} = (2p_1 + p_2)$ $\hat{z}_{12}/3 = \hat{z}_4(2p_1 + p_2)4d_4$ or $v_{31} = v_4 = 0$ if $d_{31} \neq 0$ and $v_4 = 0$ was studied above. If $d_{31} = 0$ then $d_{32} = 0$, $v_{32} = 0$, and $\hat{z}_{31}(2p_1 + p_2) = 0$ or $v_{31} = 0$ from (2.1). Further $p_1\hat{z}_{12} = 3p_1\hat{z}_{11} + p_2\hat{z}_{12} = 0$ from (1,1) and (3,1). It follows that $3v_{11}p_1(k) = v_{12}(2p_1(k) + p_2(k))$ $k = 1,2$ or $v_{11} = v_{12} = v_0 = 0$. Finally (3,2)$i = 2$ gives $\hat{z}_4(2p_1 + p_2) = 0$, $v_4 = 0$ and $V = 0$.

B.4: Case $d_4 \neq 0$, $d_{4i} \neq 0$ $i = 1,2$ with the choice $v_{42} = ... = 0$ but $v_4 \neq 0$, $v_{41} \neq 0...$.

Theorem 6: The only possible solutions are the trivial $V = W = Z = 0$ (Table 2). From (5,3)$i = 1,2$, (4,2)$i = 1,2$ we get $(p_2 - p_1)\hat{z}_4 = (p_2 - p_1)\hat{z}_{41} = p_2\hat{z} = 0$, $dp_1\hat{z}_{41} = d_{41}\hat{z}p_1$ from which we get $dv_{41} = 0$, $v = 0$. Further from (5,2), (4,1)$i = 2$ we get $p_2\hat{z}_{32} = 0$ $d_{31}p_1\hat{z}_{41} = d_{41}p_1\hat{z}_{31}$, $(2p_2 + p_1)\hat{z}_{12} = 0$.

B.4.1.: If $d = 0$ from (4,1)$i = 1$ and (1) we find $(2p_1 + p_2)(\hat{z}_{11}d_{41} - \hat{z}_{41}) = (p_1 - p_2)(\hat{z}_{11} - \hat{z}_{21}) = 0$ or $(2p_1 + p_2)\hat{z}_{11} = 3A_1v_{41}/(p_2 - p_1)d_{41}$, $(p_1 - p_2)\hat{z}_{11} = 3v_{12}A_1/(2p_2 + p_1)$, $A_i = q_ip_j - q_jp_i$. We deduce $v_{11}d_{41} = v_{41}$ and $v_{12} = 0$. Then from $p_2\hat{z}_0 = 0$, $p_1(\hat{z}_0 - \hat{z}_{11}) = 0$ we get $v_0 = v_{11} = v_{41} = 0$. Finally with (3,2), (3,1), (2,1) we get $\hat{z}_4p_i = 0$, $\hat{z}_{3i}p_1 = \hat{z}_{3i}p_2 = 0$ or $v_4 = v_{3i} = 0$ and $V = 0$.

B.4.2.: If $d \neq 0$, $v_{41} = 0$ then $(2p_i + p_j)\hat{z}_{1i} = 0$ and from (2,1), (3,1): $\hat{z}_{1i}p_i$ $(3d_{4i} - d_{3i}) + d_{4i}p_j\hat{z}_{1j} = \hat{z}_{3i}p_j = d_{3i}(2p_i\hat{z}_{1i} + p_j\hat{z}_{1j})$ and we deduce $(d_{3i} - d_{4i})$ $v_{1j} = 0$, $v_{1i}(2d - 3(d_{3i} - d_{4i})) = 0$. (i) If $d_{3i} = d_{4i}$ then $v_{1i} = 0 \longrightarrow v_0 = v_{3i} = 0$ and (3,2) $p_j\hat{z}_4 = 0$ gives $v_4 = 0$ and $V = 0$. (ii) If $d_{3i} \neq d_{4i}$, $v_{1j} = 0 \longrightarrow v_0 = 0$ and the same $V = 0$ result. (iii) If $d_{31} = d_{41}$, $d_{32} \neq d_{42} \longrightarrow v_{11} = 0$ then $\hat{z}_{31}p_2 = d_{41}p_2\hat{z}_{12}$ gives $v_{12} = v_0 = v_{31} = 0$, $\hat{z}_{32}p_1 = 0 \longrightarrow v_{32} = 0$ and still $V = 0$.

B.5.: Case $d_4 \neq 0$, $d_{4i} = 0$, $i = 1,2$, Table 3 with representation II $v_0 = \Sigma v_{1i}$, $z_0 = \Sigma z_{1i}$, $w_0 = \Sigma w_{1i}$ but V, Z, W can have the same monomials as D. We define $A_i = q_ip_j - q_ip_j$, $\bar{A} = \Sigma v_{1i}A_i/\ p_i$, $c = d_{31} + d_{32} + 1 - d$, $\bar{p} = p_1 + p_2$ and find two fundamental relations

$$\bar{A}(d_{31}d_{32} - d_4 - (d-2)c) = 0,$$
$$\Sigma d_{3j}\hat{z}_{3i}p_j = \bar{A}(d_4 - c) = \bar{A}(d_4 - c)d_{31}d_{32}/d_4 \qquad (B.7)$$

hence three cases $\bar{A} = 0$, $d_{31}d_{32} = d_4$, $d_4 = c$.

B.5.1.: If $\bar{A} = 0$ then $v_{11} = v_{12} = v_0 = 0$ (which violates assumption 2) and from (1,1), (2,1), (2,2), (3,2), (4,3) get $\hat{z}\bar{p} = \hat{z}p_j + \hat{z}_{3i}p_i = 0 \longrightarrow p_i(\hat{z} - \hat{z}_{3i}) = 0$ $\longrightarrow v = v_{3i}$; $\hat{z}_4\bar{p} = p_i\hat{z}(d - d_{3i}) + \hat{z}_4p_j = 0 \longrightarrow v_4 = (d - d_3)v$, $d_{31} = d_{32} = d_3$; $d_4\hat{z}p_j = d_3\hat{z}_4p_j \longrightarrow d_4v = d_3v_4$, $d_4 = d_3(d - d_3)$. We obtain a similarity solution

$$V = vu_1u_2(1 + \Sigma u_2 + (d - d_3)u_1u_2), \ Dv = (1 + d_3u_1u_2)V,$$
$$\mathcal{V} = vu_1u_2/(1 + d_3u_1u_2), \ \mathcal{Z} = -\bar{q}\mathcal{V}/\bar{p}(1), \ \mathcal{W} = -\bar{q}\mathcal{V}/\bar{p}(2) \qquad (B.8)$$

B.5.2.: If $d_4 = d_{31}d_{32}$, $\bar{A} \neq 0$, $(d-2)c = 0$ then either $c = 0$ or $c \neq 0$. We begin with the $c = 0$ case. From $\Sigma\hat{z}_{3i}p_id_{3j} = \Sigma\hat{z}_{3i} = 0$ we find two subcases depending whether $d_{31} = d_{32}$ or not. (i) If $d_{31} \neq d_{32}$ then $\hat{z}_{3i}p_i = 0$ or $p_j\hat{z}_{3i} = A_jv_{3i}/p_i$ while (3,1) gives $p_j\hat{z}_{3i} = A_id_{3i}v_{1i}/p_j$ leading to $v_{3i} = v_{1i} = 0$ or to the previous $\bar{A} = 0$ case. (ii) If $d_{31} = d_{32} = d_3$, from (2,2), (4,3), (3,1) we get $z_4\bar{p} = -v_4\bar{q} + \bar{A}d_4$, $z_{3i}\bar{p} = -q_jv_{3i} + A_jv_4d_3/d_4p_j + d_3\bar{A} = -v_{3i}\bar{q} + d_3v_{1i}A_i/p_j$ or $p_i(v_{3i} - v_4d_3/d_4) = d_3p_jv_{1j} \longrightarrow v_{1j} = 0$ and still the $\vec{A} = 0$ case.
We go on with the $d = 2$, $c = d_{31} + d_{32} - 1 \neq 0$ case. From (1,1) $\hat{z}\bar{p} = \bar{A}$ and $\hat{z}\Sigma p_id_{3i} = \bar{A}(2d_4 - c)$ we find $\hat{z}p_i = \bar{A}(d_{3j} - 2d_4 + c)/\ (d_{3j} - d_{3j})$. With (2,1), (3,1), (4,3) we deduce $p_i\hat{z}_{3i} = d_{3i}\bar{A} - p_j\hat{z}$, $p_j\hat{z}_{3i} = -\ d_{3i}v_{1j}A_j/p_i + p_j\hat{z}$, $p_j\hat{z}_4 = d_{3j}p_j\hat{z}_{3i}$ and substitute into (3.2) $\bar{A}(c - 2d_{3i}) + d_{3i}\hat{z}p_i + p_j\hat{z}(2 - d_{3j}) = 0$. We obtain $\bar{A}c(d_4 - c) = 0$ which correspond either to the above cases $\bar{A} = 0, c = 0$ or to the next one $d_4 = c$. If $d_{31} = d_{32} = d_3$ we still obtain a particular case $d_3 = c = d_4 = 1$ of the next case.

B.5.3.: **Case** $d_4 = c = d_{31} + d_{32} + 1 - d$, $d_{31}d_{32} = d_4(d-1)$. We firstly choose $d_{31} = d_4$, $d_{32} = d - 1 \neq d_4$ and from (1,1) and $\hat{z}\Sigma p_id_{3i} = \bar{A}d_4(d-1)$ we deduce $\hat{z}p_1 = \bar{A}(d-1)$, $\hat{z}p_2 = 0$. From (2,1), (3,1) we get $p_1\hat{z}_{31} = d_4\bar{A}$, $\hat{z}_{32}p_2 = 0$, $\hat{z}_{31}p_2 = d_4v_{12}A_1/p_1$, $\hat{z}_{32}p_1 = (d-1)v_{12}A_2/p_1$ and from (2,2), (4,3) $\hat{z}_4\bar{p} = 0$, $\hat{z}_4p_1 = d_4v_{12}A_2/p_1$ or $v_4 = 0$ and $d_4v_{12} = 0$ with two possibilities. (i) If $d_4 \neq 0$, $v_{12} = 0$ we obtain a similarity wave

$$V = v_{11}(1 + u_1 + (d-1)u_1u_2 + d_4u_1^2u_2), \ \mathcal{V} = v_{11}/(1 + u_2),$$
$$\mathcal{Z} = -q_2\mathcal{V}/p_2(1), \ \mathcal{W} = -q_2\mathcal{V}/p_2(2) \qquad (B.8)$$

(ii) If $d_4 = 0$ we still deduce $v_{12} = 0$ and another similarity wave

$$V = v_{11}(1 + u_1 + (d-1)u_1u_2), \ \mathcal{V} = v_{11}/(1 + u_2), \ \mathcal{Z} = -q_2\mathcal{V}/p_2(1) \qquad (B.9a)$$

Secondly we assume $d_{31} = d_{32} = d_3$ or $d_3 = d_4 = d - 1$ so that the denominator is:

$$D = (1 + u_1)(1 + u_2)(1 + d_4u_1u_2)$$

We obtain from (1,1), (2,1), (4,3), (3,1) $p_i(\hat{z} - \hat{z}_{3i}) = p_j(\hat{z}_4 - \hat{z}_{3i}) = 0$, $d_4v_{1i}A_i/p_j = \hat{z}_{3i}\bar{p}$ and deduce $v = v_{3i} + d_4v_{1j}$, $v_{3i} = v_4 + d_4v_{1i}$. The numerator V depends on four arbitrary parameters v_{1i}, v_4, d_4

$$V = v_4u_1u_2(1 + u_1)(1 + u_2) + \Sigma v_{1i}(1 + u_i)(1 + d_4u_1u_2) \qquad (B.9b)$$

and the solutions are either similarity waves or sums of two or three similarity waves

$$\mathcal{V} = \sum_1^2 v_{1i}/(1+u_j) + v_4 u_1 u_2/(1+d_4 u_1 u_2),$$
$$\mathcal{Z} = -\Sigma v_{1i} q_i/p_i(1)(1+u_j) - v_4 u_1 u_2 \bar{q}/\bar{p}(1)(1+d_4 u_1 u_2),$$
$$\mathcal{W} = -\Sigma v_{1i} q_i/p_i(2)(1+u_j) - v_4 u_1 u_2 \bar{q}/\bar{p}(2)(1+d_4 u_1 u_2) \qquad (B.9c)$$

Theorem 7 - *The possible solutions are similarity waves or sums of two or three similarity waves.*

APPENDIX C - $\mathcal{M}_{tt} = \mathcal{W}_{xx}$

We define $\mathcal{M} = M(u_1, u_2)/D(u_1, u_2)$, $\mathcal{W} = W(u_1, u_2)/D$ with M, W, D polynomials in $u_i = d_i \exp(\rho_i t + \gamma_i x)$; $\rho_i \gamma_j \neq \rho_j \gamma_i$ which satisfy:

$$D^2(M_{tt} - W_{xx}) + D(-2M_t D_t + 2W_x D_x - MD_{tt} + WD_{xx}) + 2(MD_t^2 - WD_x^2) = 0 \qquad (C.1)$$

Choosing $D = 1 + \Sigma u_i + du_1u_2$, $M = m_0 + \Sigma m_i u_i$, $W = w_0 + \Sigma w_i u_i$, (C.1) is a fifth-order polynomial with (m, n) for the coefficient of $u_i^m u_j^n$ and $\hat{m}_i f(\rho_i, \rho_j) = m_i f(\rho_i, \rho_j) - w_i f(\gamma_i, \gamma_j)$, $\bar{\rho} = \rho_1 + \rho_2$.

1. $d \neq 0$: We write down all relations except (2,1),
(1,0): $(\hat{m}_i - \hat{m}_0)\rho_i^2 = 0$, (3,2): $\hat{m}_i \rho_j^2 = 0$, (or $m_i \rho_j^2 = w_i \gamma_j^2$)

$$(1,1) \text{ and } (2,2): \quad (1-d)\hat{m}_0 \bar{\rho}^2 = 0, \ (\hat{m}_0 - \hat{m}_1 - \hat{m}_2)\rho_1\rho_2 = 0 \qquad (C.2)$$

(1,0) and (3,2) lead to the two cases: $(m_0 - m_1 - m_2)$ $(\rho_1\gamma_2 + \rho_2\gamma_1) = 0$. First assuming $m_0 \neq m_1 + m_2$ we get $w_0 = m_0\rho_i^2/\ \gamma_i^2$ $i = 1, 2$ and substituting into $(\hat{m}_0 - \hat{m}_1 - \hat{m}_2)$ $\rho_1\rho_2 = 0$ we find $\rho_1\gamma_2 = \rho_2\gamma_1$. Second we assume $m_0 = m_1 + m_2$ and get

$$(2,1): \quad (d-1)(\hat{m}_i(\rho_i^2 - \rho_i\rho_j) - \hat{m}_j(\rho_j^2 - 2\rho_i\rho_j)) = 0 \qquad (C.3)$$

If $d \neq 1$ then both $\hat{m}_0\bar{\rho}^2 = 0$ and (2,1) give $\hat{m}_i(\rho_i^2 \pm \rho_i\rho_j) \pm \hat{m}_j(\rho_j^2 \pm 2\rho_i\rho_j) = 0$ $\longrightarrow \hat{m}_i\rho_i^2 + \hat{m}_j 2\rho_i\rho_j = 0$. Eliminating w_i, w_j we find either $m_1 = m_2 = m_0 = 0$ or $\gamma_2\rho_1 = \gamma_1\rho_2$ which are not possible.

2. $d = 0$: Adding constants in $\mathcal{M}, \mathcal{W}$ we can put $m_2 = q_2 = 0$ and (C.1) becomes: (1,0): $\rho_1^2(\hat{m}_0 - \hat{m}_1) = 0, \rho_2^2\hat{m}_0 = 0$, (2,1): $\hat{m}_1(\rho_1 - \rho_2)^2 = 0$, (1,1): $2\hat{m}_0\rho_1\rho_2 - \hat{m}_1\rho_2^2 = 0$. Eliminating w_0, w_1 into (1,0) and (1,1) we find $m_0 B/\gamma_2^2 = m_1(B - 2\rho_1\gamma_1)/(\gamma_1 - \gamma_2)^2$, $m_0 2\rho_2/\gamma_2 = m_1(-B + 2\gamma_2\rho_2)/\ (\gamma_1 - \gamma_2)^2$ with $B = \gamma_1\rho_2 + \ \gamma_2\rho_1$. We find either $m_0 = m_1 = \mathcal{M} = 0$ or $\gamma_2\rho_1 = \gamma_1\rho_2$ which are not possible.

3. $d = 1$: We find from $m_0 = m_1 + m_2$ and (C.2-3) the sum of two similarity waves: $M = \Sigma m_i(1 + u_i)$, $W = \Sigma m_i(\rho_i/\gamma_i)^2\ (1 + u_i)$, $D = (1 + u_1)(1 + u_2)$, $\mathcal{M} = \Sigma m_i/(1 + u_j)$, $\mathcal{W} m_i\rho_i^2/\gamma_i^2(1 + u_j)$.

APPENDIX D - 5 DENSITIES AND 3 CONSERVATION LAWS

For 5 densities $\mathcal{N}_i$, $\mathcal{M}_i$, $i = 1,2$, $\mathcal{M}_4$ satisfying three conservation laws

$$\partial_t \mathcal{M}_2/2 + (\partial_t + \partial_x)(\mathcal{N}_2 + \mathcal{M}_1/4) = 0, \ \partial_t \mathcal{M}_2/2 + (\partial_t - \partial_x)(\mathcal{N}_1 + \mathcal{M}_4/4) = 0 \quad (D.1)$$

$$(\partial_t - \partial_x)\mathcal{N}_1 + (\partial_t + \partial_x)\mathcal{N}_2 = 0 \quad (D.2)$$

we assume properties 1, 2, 3 of section 2 with a common fourth-order denominator D of the type (B.1). For the 3 densities $\mathcal{V} = \mathcal{M}_2$, $\mathcal{Z} = \mathcal{N}_2 + \mathcal{M}_1/4$, $\mathcal{W} = \mathcal{N}_1 + \mathcal{M}_4/4$ satisfying the 2 conservations laws (D.1) we can apply the results of section 2 (with $q_i = \rho_i$, $p_i(1) = \rho_i + \gamma_i$, $p_i(2) = \rho_i - \gamma_i$ everywhere) and for (1+1)-dimensional solutions we know that D is factorized $D = (1+u_1)(1+u_2)\,(1+d_4 u_1 u_2)$. Substituting such D into (D.2), what are the possible rational solutions ?

More generally going back to 2 densities but with only one conservation law,

$$\mathcal{V}_t + \mathcal{V}_x + \alpha \mathcal{Z}_t + \beta \mathcal{Z}_x = 0 \quad (D.3)$$

further assuming factorized D, are the possible (1+1)-dimensional rational solutions only sums of similarity waves ? Two cases occur: D of the second or of the fourth order. We apply representation II with Tables 1 and 3. The main difference with Appendix B is that here we only have one linear relation (D.3).

D.1: $D = (1+u_1)(1+u_2)$ with v_{1i}, v, z_{1i}, z as arbitrary parameters (Table 1). We have $d = 1$ and get $p_j \hat{z} = 0$, $j = 1,2$ from (2,1), or $v = z = 0$. Consequently from $\hat{z}_{1i} p_j = 0$ we find $\mathcal{V} = \Sigma v_{1i}/(1+u_j)$, $\mathcal{Z} = -\Sigma v_{1i} q_j/\ p_j(1+u_j)$ and the solutions are sums of two similarity waves.

D.2: $D = (1+u_1)(1+u_2)(1+d_4 u_1 u_2)$ with v_{1i}, v, v_{3i}, v_4; z_{1i}, z, z_{3i}, z_4 as arbitrary parameters (Table 3); $d = 1 + d_4$, $d_4 = d_{3i} = c$. The discussion is similar to the Appendix B.5 one. From $\bar{p}\hat{z}_4 = 0$ we find $z_4 p_j = v_4 A_j/\bar{p}$. From (1,1), (2,1) we get $\bar{A} d_4 = \hat{z} p_j + \hat{z}_{3i} p_i = \hat{z}\bar{p}$, $p_i(\hat{z} - \hat{z}_{3i}) = 0$ while (3,2), (1,1) give $d_4 v_{1j} A_j/p_i = p_j(\hat{z} - \hat{z}_{3i}) \longrightarrow v = v_{3i} - d_4 v_{1j}$, $z = z_{3i} - d_4 z_{1j}$, $z_{1j} = -q_1 v_{1j}/p_j$. Using (4,3) $\hat{z}_{3i} p_j = \hat{z}_4 p_j$ and eliminating z_{3i} into (3,2) $d_4 v_{1i} A_i/p_j = \hat{z}_{3i}\bar{p}$ we get $v_4 = v_{3i} - d_4 v_{1i}$, $z_4 = z_{3i} - d_4 z_{1i} = -v_4 \bar{q}/\bar{p}$. V depends on 4 parameters v_{1i}, v_4, d_4

$$V = \sum_1^2 v_{1i}(1+u_i)(1+d_4 u_1 u_2) + v_4 u_1 u_2 (1+u_1)(1+u_2) \quad (D.4)$$

while $\mathcal{V}$ and $\mathcal{Z}$ are the sums of 2 or 3 similarity waves and are written down in (B.9c).

In conclusion, even if two independent densities satisfy only one conservation law, but provided they are associated (via our assumptions 1, 2, 3) to 3 other ones with two conservation laws, then still only sums of similarity waves can occur for the 5 densities.

TABLES

As illustration we give the details for the construction of Table 1 so that the reader can extend the method to the other Tables. Starting with the first eq. (2.1a): $0=0_+V+0_k Z$, $k=1$, $0_+=\partial_t+\partial_x$, $0_k=\alpha_k\partial_t+\beta_k\partial_x$, $V=v_{00}+V/D$, $Z=z_{00}+Z/D$ we get:

$$Z0_1D+V0_+D=D(0_1Z+0_+V) \qquad \text{(T1)}$$

which is eq. (2.7a). V,Z,D are polynomials (see Table1) in the variable $u_i=\exp(\rho_i t+\gamma_i x)$, $i=1,2$ and the operators $0_+,0_k$ applied to u_i introduce new parameters: $p_j(k)=\alpha_k\rho_j+\beta_k\gamma_j$, $q_j=\rho_j+\gamma_j$, $p_i(k)q_j-p_j(k)q_i=(\rho_i\gamma_j-\rho_j\gamma_i)(\alpha_k-\beta_k)\neq 0$. Here k is always 1 and for simplicity we write $p_i(1)=p_i$. We find $0_1D=\Sigma u_ip_i+u_i^2u_j(2p_i+p_j)+du_1u_2(p_1+p_2)$ while for 0_+D we change $p_i\to q_i$. In order to simplify the notations let us, for any couple z_{mi},v_{mi} define $\hat{z}_{mi}f(p_i,p_j)=z_{mi}f(p_i,p_j)+v_{mi}f(q_i,q_j)$. We get $0_1Z=\Sigma u_iz_{1i}p_i+u_i^2u_jz_{3i}(2p_i+p_j)+u_1u_2z(p_1+p_2)$, change $p_i,z_{mi}\to q_i,v_{mi}$ for 0_+V and $z_{mi}\to\hat{z}_{mi}$ for 0_1Z+0_+V. (T1) becomes:

$$[\Sigma z_{1i}(1+u_i)+z_{3i}u_i^2u_j+zu_1u_2][\Sigma u_ip_i+u_i^2u_j(2p_i+p_j)d_{3i}+u_1u_2d\bar{p}]+v_{mi},q_i$$
$$=[1+\Sigma u_i+du_1u_2+\Sigma d_{3i}u_i^2u_j][u_1u_2\hat{z}\bar{p}+\Sigma u_ip_i\hat{z}_{1i}+u_i^2u_j\hat{z}_{3i}(2p_i+p_j)] \qquad \text{(T2)}$$

with $\bar{p}=p_1+p_2$. The coefficients, called (m,n), of $u_i^mu_j^n$ are zero:

(1,0) or u_i: $\hat{z}_{1i}p_j=0\to z_{1i}=-v_{1i}q_j/p_j\to\hat{z}_{1i}p_i=A_iv_{1i}/p_j$ with $A_i=q_ip_j-q_jp_i\neq 0$. (1,1) or u_1u_2: $\hat{z}\bar{p}=(d-1)\Sigma p_i\hat{z}_{1i}=(d-1)\bar{A}$ with $\bar{A}=\Sigma v_{1i}A_i/p_j$, $j\neq i$. (2,2) or $u_1^2u_2^2$: $\Sigma\hat{z}_{3i}p_i=0$ (3,1) or $u_i^3u_j$: $d_{3i}\hat{z}_{1i}p_i=\bar{p}\hat{z}_{3i}$ (2,1) or $u_i^2u_j$: $d_{3i}(\bar{A}+z_{1i}p_i)=p_j\hat{z}+\hat{z}_{3i}(2p_i+p_j)$ or with (3,1) $d_{3i}\bar{A}=\hat{z}p_j+\hat{z}_{3i}p_i\to(d_{31}+d_{32})\bar{A}=\hat{z}\bar{p}+\Sigma\hat{z}_{3i}p_i$ or with (1,1),(2,2) $\bar{A}(d_{31}+d_{32}+1-d)=0$, $2d_{31}d_{32}\bar{A}=\hat{z}\Sigma p_id_{3i}+\Sigma p_i\hat{z}_{3i}d_{3j}$. (3,2) or $u_i^3u_j^2$: $\hat{z}d_{3i}p_i=d\hat{z}_{3i}p_i$ and $\hat{z}\Sigma d_{3i}p_i=0$ with (2,2) and $2d_{31}d_{32}\bar{A}=\Sigma p_i\hat{z}_{3i}d_{3j}$ with the last (2,1) relation. (3,3) or $u_i^3u_j^3$: $\Sigma\hat{z}_{3i}d_{3j}(p_j-p_i)=0$. Multiplying (3,1) by d_{3j} and summing up over i we get: $d_{31}d_{32}\bar{A}=\Sigma p_i(\hat{z}_{3i}d_{3j}+\hat{z}_{3j}d_{3i})=2\Sigma p_iz_{3i}d_{3j}$ with (3,3). From this relation and the last (3,2) one we find finally $d_{31}d_{32}\bar{A}=0$ which is the starting point for the study of the possible solutions.

The second (2.1a) relation is the same as (T.1) with k=2 and $Z\to W$. All the Table 1 results are valid with the changes: $p_i=p_i(2)$ and $\hat{z}_{mi}\to\hat{w}_{mi}$ for $\hat{w}_{mi}f(p_i,p_j)=w_{mi}f(p_i,p_j)+v_{mi}f(q_i,q_j)$.

TABLES 1-2-3: 3 Densities and 2 Conservation Laws

Def. Tables 1-2-3: $D=1+\Sigma u_i+u_i^2u_j(d_{3i}+d_{4i}u_i)+du_1u_2+d_4u_1^2u_2^2$, $\bar{p}=p_1+p_2$, $\hat{z}_{mi}(ap_i+bp_j)=z_{mi}(ap_i(1)+bp_j(1))+v_{mi}(aq_i+bq_j)$; Tables 1,3: Repretation II, $V=\Sigma v_{1i}(1+u_i)+v_{3i}u_i^2u_j+vu_1u_2+v_4u_1^2u_2^2$, $Z=\Sigma z_{1i}(1+u_i)+z_{3i}u_i^2u_j+vu_1u_2+v_4u_1^2u_2^2$, $A_i=q_ip_j-q_jp_i\neq 0$, $\bar{A}=\Sigma_i v_{1i}A_i/p_j$ $j\neq i$; Table 2: Representation I, $V=v_0+\Sigma v_{1i}u_i+u_i^2u_j(v_{3i}+v_{4i}u_i)+vu_1u_2+v_4u_1^2u_2^2$, $Z=z_0+\Sigma z_{1i}u_i+u_i^2u_j(z_{3i}+z_{4i}u_i)+zu_1u_2+z_4u_1^2u_2^2$.

TABLE 1: $d_4=d_{4i}=v_4=z_4=0$

(1,0) $\hat{z}_{1i}p_j=0$, $\hat{z}_{1i}p_i=v_{1i}A_i/p_j$, (1.1) $\hat{z}\bar{p}=(d-1)\bar{A}$, (2,1) $d_{3i}\bar{A}=p_j\hat{z}+p_i\hat{z}_{3i}$, $2d_{31}d_{32}\bar{A}=\Sigma\hat{z}_{3i}p_id_{3j}$, $\bar{A}(d_{31}+d_{32}+1-d)=0$, (3,1) $d_{3i}v_{1i}A_i/p_j=\bar{p}\hat{z}_{3i}$, $d_{31}d_{32}\bar{A}=2\Sigma\hat{z}_{3i}p_id_{3j}$, (2,2) $\Sigma p_i\hat{z}_{3i}=0$, (3,2) $d_{3i}\hat{z}p_i=dp_i\hat{z}_{3i}$, (3,3) $\Sigma p_i(d_{3i}\hat{z}_{3j}-d_{3j}\hat{z}_{3i})=0$

TABLE 2:

(4,4) $\Sigma d_{4i}(p_i-p_j)\hat{z}_{4j}=0$, (5,3) $(p_j-p_i)(d_4\hat{z}_{4i}-d_{4i}\hat{z}_4)=0$, (5,2) $d_{3i}p_i\hat{z}_{4i}=d_{4i}\hat{z}_{3i}p_i$, (4,3) $(2p_i-p_j)(d_{4i}\hat{z}_{3j}-d_{3j}\hat{z}_{4i})=p_j(\hat{z}_4d_{3i}-d_4\hat{z}_{3i})$ (4,2) $dp_i\hat{z}_{4i}=d_{4i}\hat{z}p_i$, (3,3) $\Sigma d_{3i}(p_j-p_i)z_{3j}=\bar{p}(d_4\hat{z}-d\hat{z}_4)$, (4,1) $(\bar{p}+p_i)\hat{z}_{1i}d_{4i}=(\bar{p}+p_i)\hat{z}_{4i}$, (3,2) $d_{3i}\hat{z}p_i+d_4(\bar{p}+p_j)\hat{z}_{1i}+3d_{4i}p_i\hat{z}_{1j}=dp_i\hat{z}_{3i}+3p_i\hat{z}_{4i}+\hat{z}_4(\bar{p}+p_j)$, (2,2) $d_4\hat{z}_0\bar{p}+\Sigma p_i\hat{z}_{1j}d_{3j}=\hat{z}_4\bar{p}+\Sigma p_i\hat{z}_{3i}$, (3,1) $d_{3i}\bar{p}\hat{z}_{1i}+d_{4i}(\bar{p}+2p_i)\hat{z}_0=\hat{z}_{3i}\bar{p}+\hat{z}_{4i}(\bar{p}+2p_i)$, (2,1) $d_{3i}(\bar{p}+p_i)\hat{z}_0+d\hat{z}_{1i}p_j=\hat{z}_{3i}(\bar{p}+p_i)+\hat{z}p_j$, (1,1) $d\bar{p}\hat{z}_0=\bar{p}\hat{z}+(p_1-p_2)(\hat{z}_{11}-\hat{z}_{12})$, (1,0) $p_i\hat{z}_{1i}=p_i\hat{z}_0$ or $p_iz_{1i}+q_iv_{1i}=p_iz_0+q_iv_0$, Σ means $\sum_i$

TABLE 3: $d_{4i}=0$, $d_4\neq 0$, $v_4\neq 0$, $z_4\neq 0$

(1,0) $\hat{z}_{1i}p_j=0$, $z_{1i}p_i=v_{1i}A_i/p_j$, (1,1) $\hat{z}\bar{p}=(d-1)\bar{A}$, (2,1) $d_{3i}\bar{A}=\hat{z}p_j+\hat{z}_{3i}p_i$, (2,2) $\hat{z}_4\bar{p}=(d_4-c)\bar{A}$, $c=d_{31}+d_{32}+1-d$, (3,1) $\hat{z}_{3i}\bar{p}=d_{3i}v_{1i}A_i/p_j$, (4,3) $d_4\hat{z}_{3i}p_j=d_{3i}\hat{z}_4p_j$, (3,2) $d\hat{z}_{3i}p_i+\hat{z}_4(\bar{p}+p_j)=d_{3i}\hat{z}p_i+d_4v_{1i}A_i/p_j$, (3,3) $\Sigma\hat{z}_{3i}d_{3i}(p_j-p_i)=\bar{A}(d_4-dc)$, $d_4-d_{31}d_{32}+c(d-2)=0$, $\Sigma\hat{z}_{3i}p_jd_{3j}=\bar{A}(d_4-c)=\bar{A}(d_4-c)d_{31}d_{32}/d_4$, $\Sigma\hat{z}_{3i}p_id_{3j}=\bar{A}(d-1)c$, $\Sigma\hat{z}_{3i}p_i=\bar{A}c$, $\hat{z}\Sigma p_id_{3i}=\bar{A}(2d_4+c(d-3))$

TABLE 4: 2 Conservation Laws Representation II

$$D=1+du_1u_2+\Sigma u_i(1+d_{3i}u_iu_j)$$

(1): $(V/D)_x+(M/D)_t=0$, $V=\Sigma v_{1i}(1+u_i)+v_{3i}u_i^2u_j+vu_1u_2$, $M=mu_1u_2+\Sigma m_{1i}(1+u_i)+m_{3i}u_i^2u_j$, def. $\hat{m}_{ki}f(\rho_i,\rho_j)=m_{ki}f(\rho_i,\rho_j)+v_{ki}f(\gamma_i,\gamma_j)$, $A_i(1)=\gamma_i\rho_j-\gamma_j\rho_i\neq 0$, $\bar{A}(1)=\Sigma v_{1i}A_i(1)/\rho_j$, $\bar{\rho}=\rho_1+\rho_2$

(1,0) $\hat{m}_{1i}\rho_j=0$, $\hat{m}_{1i}\rho_i=v_{1i}A_i(1)/\rho_j$, (1,1) $\bar{\rho}\hat{m}=(d-1)\bar{A}(1)$, (2,1) $d_{3i}\bar{A}(1)=\rho_j\hat{m}+\rho_i\hat{m}_{3i}$, $2d_{31}d_{32}\bar{A}(1)=\Sigma\rho_i\hat{m}_{3i}d_{3j}$, $\bar{A}(1)(d_{31}+d_{32}+1-d)=0$, (3,1) $\bar{\rho}\hat{m}_{3i}=d_{3i}v_{1i}A_i(1)/\rho_j$, $d_{31}d_{32}\bar{A}(1)=\Sigma\rho_i\hat{m}_{3i}d_{3j}$, (3,2) $d_{3i}\hat{m}\rho_i=d\rho_i\hat{m}_{3i}$, (2,2) $\Sigma\rho_i\hat{m}_{3i}=0$, (3,3) $\Sigma\rho_i(d_{3i}\hat{m}_{3j}-d_{3j}\hat{m}_{3i})=0$

(2): $(V/D)_t+(W/D)_x=0$, same V, same D, $W=\Sigma w_{1i}(1+u_i)+w_{3i}u_i^2u_j+wu_1u_2$, $\hat{w}_{ki}f(\gamma_i,\gamma_j)=w_{ki}f(\gamma_i,\gamma_j)+v_{ki}f(\rho_i,\rho_j)$, $A_i(2)=-A_i(1)$, $\bar{\gamma}=\Sigma\gamma_i$, $\bar{A}(2)=\Sigma v_{1i}A_i(2)/\gamma_j$

(1,0) $\hat{w}_{1i}\gamma_j=0$, $\hat{w}_{1i}\gamma_i=v_{1i}A_i(2)/\gamma_j$, (1,1) $\bar{\gamma}\hat{w}=(d-1)\bar{A}(2)$, (2,1) $d_{3i}\bar{A}(2)=\gamma_j\hat{w}+\gamma_i\hat{w}_{3i}$, $2d_{31}d_{32}\bar{A}(2)=\Sigma\gamma_i\hat{w}_{3i}d_{3j}$, $\bar{A}(2)(d_{31}+d_{32}+1-d)=0$, (3,1) $\bar{\gamma}\hat{w}_{3i}=d_{3i}v_{1i}A_i(2)/\gamma_j$, $d_{31}d_{32}\bar{A}(2)=\Sigma\gamma_i\hat{w}_{3i}d_{3j}$, (3,2) $d_{3i}\hat{w}\gamma_i=d\gamma_i\hat{w}_{3i}$, (2,2) $\Sigma\gamma_i\hat{w}_{3i}=0$, (3,3) $\Sigma\gamma_i(d_{3i}\hat{w}_{3j}-d_{3j}\hat{w}_{3i})=0$

Are Heavy Higgs Interactions Observable at the 100 GeV Scale?*

*G. Gounaris**, F.M. Renard, and D. Schildknecht*

Institut für Theoretische Physik der Universität Bielefeld,
D-4800 Bielefeld 1, Fed. Rep. of Germany and
Laboratoire de Physique Mathématique***, USTL,
F-34095 Montpellier Cedex 2, France

* Supported by the program PROCOPE
for French-German scientific collaboration
** Supported by Deutsche Forschungsgemeinschaft.
Permanent address: Department of Theoretical Physics,
University of Thessaloniki, Thessaloniki, Greece
*** Unité de Recherche Associée au CNRS, URA 768

1. Introduction

The standard $SU(2)_L \times U(1)_Y$ electroweak theory /Glashow, 1961; Weinberg, 1967; Salam, 1968/ has been remarkably successful in quantitatively predicting new phenomena. The Higgs sector of the theory, however, which is essential for the perturbative renormalizability /t' Hooft, 1971/ of the theory beyond the one-loop order /Veltman, 1989/, has remained empirically untested /Gunion et al., 1989/ so far.

In fact, there have been almost no stringent limits on the value of the mass of the Higgs-boson, m_H. Experiments on e^+e^- annihilation at LEP 100 and LEP 200 /Proc. Workshop "Polarization at LEP", 1988; Proc. Workshop on LEP Physics, 1989/ will be able to produce and to identify a Higgs particle, provided its mass is less than the mass of the $W^\pm$ boson, M_W. Indirectly, via the dependence on m_H of the radiative corrections to the Born-term formula for the Z^0 mass, the high-precision data at LEP 100 will eventually also set limits on m_H, in particular, one might be able to draw conclusions on whether $m_H \cong M_W$ or rather $m_H \gtrsim 0.5$ TeV. Here, it is assumed, of course, that a reliable value for the top-quark mass will eventually be available.

The present paper is concerned with the possibility of a very large value of the mass of the Higgs boson /Veltman, 1977/, i.e., to be definite, it will be assumed that future experiments will have ruled out the existence of a Higgs scalar below a mass of the order of 1 TeV.

The quadratic divergence in the mass of the Higgs scalar indeed develops a tendency to drive the Higgs mass upwards. Unless one postulates an extension of the $SU(2)_L \times U(1)_Y$ theory by, e.g., introducing supersymmetry /Haber et al., 1985/, in order to protect a low Higgs-boson mass (thereby satisfying "naturalness" /t' Hooft, 1980/), one may in fact be led to consider the limit of $m_H \to \infty$ as a realistic possibility. In this (strongly interacting) limit, by invoking an analogy of the $SU(2)_L \times U(1)_Y$ electroweak theory to the non-linear sigma model, the existence of new gauge bosons of masses larger than the masses of the $W^{\pm}$ and Z^0 has been conjectured /Casalbuoni et al., 1985, 1987, 1988; Rosenfeld et al., 1988/. A large mass, m_H, i.e., $m_H \gg v(\text{Higgs}) \cong 250$ GeV, where v(Higgs) denotes the vacuum-expectation value of the Higgs field, would indicate an analogy of the role of the scalar sector within the electroweak theory to the way chiral $SU(2) \times SU(2)$ symmetry is broken in QCD, where $M(\text{chiral}) \gtrsim 1$ GeV $\gg v(\text{chiral}) = 0.09$ GeV.

It is well-known, however, that the limit of $m_H \gtrsim 1$ TeV is a problematic one within the $SU(2)_L \times U(1)_Y$ electroweak theory. Partial wave unitarity applied to the tree amplitudes of longitudinal vector boson scattering /Lee et al., 1979/, improved by employing the renormalization-group to introduce a running scalar-field self-coupling $\lambda(\mu)$ /Maiani et al., 1978; Cabibbo et al., 1979/ leads to /Marciano et al., 1989/

$$\lambda(\mu) = \frac{G_F}{\sqrt{2}} \frac{m_H^2}{1 - \frac{3}{2\pi^2} \frac{G_F}{\sqrt{2}} m_H^2 \ln \frac{\Lambda}{m_H}} \leq \frac{8\pi}{5} \quad , \tag{1.1}$$

which implies

$$m_H^2 \leq \frac{2\pi^2\sqrt{2}}{3G_F} \left[\ln \frac{\Lambda}{m_H} + \frac{5\pi}{12}\right]^{-1} \quad , \tag{1.2}$$

i.e., for a given value of $m_H \lesssim 780$ GeV, there exists a scale $\Lambda > m_H$, such that perturbative partial wave unitarity is violated at energies beyond this scale. For $\Lambda = 2m_H$, one finds $m_H \lesssim 630$ GeV, while identification of Λ with the extremely large value of the Planck scale requires $m_H \lesssim 140$ GeV. It is worth emphasizing that the perturbative bound (1.2) coincides with the bound on m_H which has been found /Kuti, 1988; Lüscher et al., 1988/ by employing non-perturbative lattice methods. It may thus be considered as fairly well established that the standard $SU(2)_L \times U(1)_Y$ theory cannot be literally true, or has to be modified in one way or another, unless experiments will reveal the existence of a Higgs-scalar particle below an upper bound of the order of 1 TeV.

The above conclusion on an upper limit of the allowed values of m_H is confirmed /Gounaris et al., 1989/ by an analysis of the Higgs-boson propagator at the (iterated) one-loop order (compare Fig. 1.1) which will be

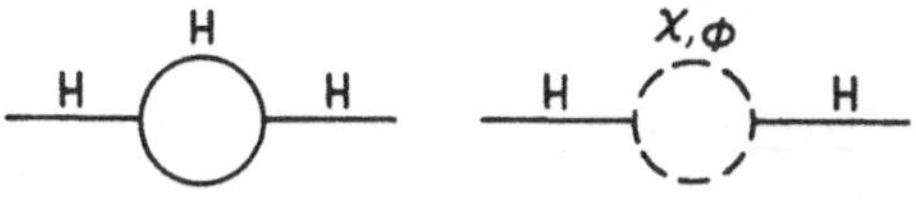

Figure 1.1: The one-loop corrected Higgs propagator (Tadpole-diagrams are not important within the present context).

presented below. We will see that in the limit of m_H becoming large, the Higgs propagator develops a pole in the spacelike region, which violates locality and causality requirements and is thus unacceptable on physical grounds. It is found, however, that the imaginary part of the propagator remains reasonable for any value of m_H, thus suggesting to tentatively keep it, while appropriately modifying /Gounaris et al., 1989/ the real part of the propagator in such a manner that the offending pole in the spacelike region does not occur. Technically, this is achieved by introducing a dispersion relation defining the new non-standard modified Higgs propagator in terms of the imaginary part of the standard Higgs propagator. We note that such a modification of the Higgs sector of the electroweak theory is an unavoidable minimal one, if at all one wants to keep the notion of the Higgs mechanism in the limit of $m_H \gtrsim 1$ TeV.

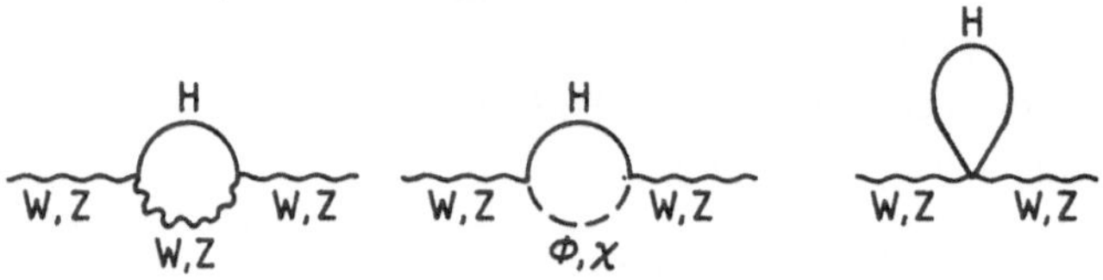

Figure 1.2: Higgs-boson contribution to the radiative correction, Δr.

The modified propagator will subsequently be applied to study the radiative correction, Δr, to the Born-term formula for the Z^0 mass in the limit of $m_H \gtrsim 1$ TeV by replacing the elementary Higgs propagator in Fig. 1.2 by the modified one.

The scope of, or the precise limitations of our attempt to modify the $SU(2)_L \times U(1)_Y$ theory in such a manner that the limit of $m_H \gtrsim 1$ TeV does not produce an obvious defect of the theory needs further investigations. Our procedure is based on a fairly rudimental use of analyticity, which nevertheless allows us to devote this paper to André MARTIN at the occasion of his 60th birthday, hoping that he will accept it, at least as an example on how vulgarized analyticity can become.

In Section 2, we will present our analysis of the Higgs-boson propagator, in particular, we will show, how the additional pole in the spacelike region comes about, and we will introduce and discuss the modified propagator in some detail. In Section 3, we will present the consequences of the modified propagator for the radiative correction, Δr. Some concluding remarks will be given in Section 4.

2. The Higgs-Boson Propagator and its Modified Form

For $m_H >> M_W$, the dominant vacuum-polarization contributions to the Higgs propagator are due to the "would-be-Goldstone-boson" loops corresponding to the longitudinal degrees of freedom of the $W^{\pm}$ and Z^0 bosons, and the Higgs-boson loop (compare Fig. 1.1). The Higgs-field couples to the longitudinal vector bosons with the strength

$$\lambda(m_H) = G_F \frac{m_H^2}{\sqrt{2}} \quad , \tag{2.1}$$

which becomes large, $\lambda >> 1$, for sufficiently large m_H, implying the transition to strong interactions. After mass renormalization, the inverse Higgs propagator in an arbitrary ξ gauge ($\xi \geq 0$) is given by

$$\begin{aligned} \tilde{\Delta}^{-1}_{H,\xi}(p^2) = {} & p^2 - m_H^2 \\ & - \frac{3\sqrt{2}}{32\pi^2} G_F m_H^4 \left[\left(1 - \frac{4\xi M^2}{p^2}\right)^{1/2} \ln\left(\frac{\left(1 - \frac{4\xi M^2}{p^2}\right)^{1/2} + 1}{\left(1 - \frac{4\xi M^2}{p^2}\right)^{1/2} - 1}\right) \right. \\ & \left. - \left(1 - \frac{4\xi M^2}{m_H^2}\right)^{1/2} \ln\left(\frac{1 + \left(1 - \frac{4\xi M^2}{m_H^2}\right)^{1/2}}{1 - \left(1 - \frac{4\xi M^2}{m_H^2}\right)^{1/2}}\right) + H(p^2)\right] , \end{aligned} \tag{2.2}$$

where $H(p^2)$ denotes the (ξ-independent) contribution of the Higgs-boson loop in Fig. 1.1,

$$H(p^2) = \begin{cases} 6 \left[\left(\frac{4m_H^2}{p^2} - 1\right)^{1/2} \operatorname{arctg} \left(\frac{4m_H^2}{p^2} - 1\right)^{-1/2} - \sqrt{3} \operatorname{arctg} \frac{1}{\sqrt{3}} \right] , & \text{for } 0 \leq \frac{p^2}{m_H^2} \leq 4 \\ 3 \left(1 - \frac{4m_H^2}{p^2}\right)^{1/2} \left[-i\pi\theta\left(\frac{p^2}{m_H^2} - 4\right) + \ln \left| \frac{1 + \left(1 - \frac{4m_H^2}{p^2}\right)^{1/2}}{1 - \left(1 - \frac{4m_H^2}{p^2}\right)^{1/2}} \right| \right] - 6\sqrt{3} \operatorname{arctg} \frac{1}{\sqrt{3}} , & \text{for } p^2 > 4m_H^2 \text{ or } p^2 < 0 . \end{cases} \tag{2.3}$$

In (2.1) the $W^{\pm}$-Z^0 mass difference has been neglected and M denotes the average mass

$$M = \frac{1}{2} (M_W + M_Z) \quad . \tag{2.4}$$

G_F denotes the Fermi coupling.

We are interested in radiative correction-effects induced by the Higgs-particle in the limit of $m_H >> M$. In this limit, we expect radiative corrections to become independent of the threshold factors in the Higgs propagator, which are due to the finite values of M_W and M_Z. Accordingly, we put $M^2 = 0$ in (2.1) and obtain

$$\tilde{\Delta}^{-1}_{H,M^2=0}(p^2) = p^2 - m_H^2 - \frac{3\sqrt{2}}{32\pi^2} G_F m_H^4 \left[\ln\left(-\frac{p^2}{m_H^2}\right) + H(p^2)\right] \quad , \tag{2.5}$$

i.e., the propagator becomes gauge invariant (ξ-independent) in this limit. We note that the expression (2.5) may alternatively be derived by treating the longitudinal vector-bosons in the unitary gauge and considering the limit of

$$|p^2|,\ m_H^2 \gg M^2 \quad , \tag{2.6}$$

which thus defines the region of validity of the form (2.5) of the (inverse) Higgs propagator.

Turning to a detailed discussion of the inverse propagator (2.5), we note that its imaginary part at $p^2 = m_H^2$ coincides with the decay width of the Higgs scalar into longitudinally polarized $W^\pm$ and Z^0 vector bosons, which for $m_H \gg M_W$ is given by

$$m_H \Gamma_H = \frac{3\sqrt{2}}{32\pi} G_F m_H^4 = \frac{3\sqrt{2}}{16\pi} \frac{\lambda^2(m_H^2)}{G_F} \quad , \tag{2.7}$$

i.e., with increasing m_H the width of the particle becomes exceedingly large, and the increase of the scalar self-coupling, $\lambda \gg 1$, corresponds to the transition to a strongly interacting situation.

Closely examining the real part of the inverse Higgs propagator in (2.5), reveals that it develops a zero for p^2 in the spacelike region, $-m_H^2 < p^2 < 0$, corresponding to a (physically unacceptable) pole of the propagator in addition to the (physical) pole at $p^2 = m_H^2$. A simple numerical analysis yields the position of this pole, denoted by p_{SP}^2, as a function of m_H. The values of p_{SP}^2/m_H^2 are given in Table 2.1. In order to study the sensitivity of the position of the pole against variation of the details of the Higgs interactions, in Table 2.1, we also give the position of the pole, for the case of a vanishing Higgs-boson-loop contribution, putting $H(p^2) = 0$ in (2.5). Looking at Table 2.1, we observe that for sufficiently large m_H, we have

$$4M^2 \ll -p_{SP}^2 \lesssim m_H^2 \lesssim \Lambda^2 \quad , \tag{2.8}$$

Table 2.1: The position, p_{SP}^2, of the pole of the Higgs propagator in the spacelike region. Case A corresponds to (2.5), while in case B, $H(p^2) = 0$.

m_H	p_{SP}^2/m_H^2 (A)	p_{SP}^2/m_H^2 (B)	$\lvert p_{SP}^2\rvert/4M^2$ (A)
2	-0.095	-0.158	13
3	-0.217	-0.377	67
4	-0.296	-0.541	162
10	-0.426	-0.886	1459
∞	-0.459	-0.999	∞

i.e., the pole lies within a region, where no such pole should appear, as the theory should at least be valid in the range of $|p^2| \lesssim m_H^2$. The existence of the pole is of course closely related to the (Landau) pole of the running coupling constant /Maiani et al., 1978; Cabibbo et al., 1979/, which according to (1.1) becomes infinite, $\lambda(\Lambda) \to \infty$, for an appropriately chosen scale Λ.

As announced in the introduction, we propose to extend the $SU(2)_L \times U(1)_Y$ theory into the region of $m_H \gtrsim 1$ TeV by conservatively modifying the interaction in the Higgs sector in such a manner that the Higgs propagator will become physically acceptable. Keeping the imaginary part of the propagator (2.5), and introducing the function

$$h(y,m_H^2) = -\mathrm{Im}\,(\tilde{\Delta}_{H,M^2=0}(y+i\,\varepsilon)) \quad , \tag{2.9}$$

we demand decent analyticity properties by constructing a modified propagator via

$$\Delta_H(p^2) = \frac{1}{\pi}\int_0^\infty \frac{h(y,m_H^2)}{p^2-y}\,dy \quad . \tag{2.10}$$

In terms of the original propagator, we obviously have

$$\Delta_H(p^2) = \tilde{\Delta}_{H,M^2=0}(p^2) - \tilde{\Delta}_{SP}(p^2) \tag{2.11}$$

where by definition

$$\tilde{\Delta}_{SP}(p^2) = \frac{R_L}{p^2 - p_{SP}^2} \quad , \tag{2.12}$$

with R_L denoting the residue of the pole of $\tilde{\Delta}_{H,M^2=0}$ at $p^2 = p_{SP}^2$,

$$R_L = [(\frac{\partial\tilde{\Delta}_H^{-1}(p^2)}{\partial p^2})_{p^2=p_{SP}^2}]^{-1} \quad . \tag{2.13}$$

While $\tilde{\Delta}_{H,M^2=0}(p^2)$ and $\tilde{\Delta}_{SP}(p^2)$ are separately singular in the spacelike region, the modified propagator, $\Delta_H(p^2)$, is obviously regular.

The propagators (2.5,10) contain contributions of the longitudinal vector bosons, as well as the contribution of the Higgs-boson loop in Fig. 1.1. One may entertain the point of view that in the limit of $m_H \gg M_{W,Z}$, when the Higgs particle has actually "degenerated" into a broad continuum of vector-boson states, the evaluation of the Higgs-boson loop may not be meaningful without, e.g., introducing form factors, typical for strong-interaction physics. Accordingly, we will actually consider two alternatively modified propagators. In case A, we will simply use the imaginary part of the full propagator (2.5) in the modified propagator (2.9,10), while in case B, we put $H(p^2) = 0$ in (2.5), which corresponds to dropping the Higgs-boson loop in Fig. 1.1.

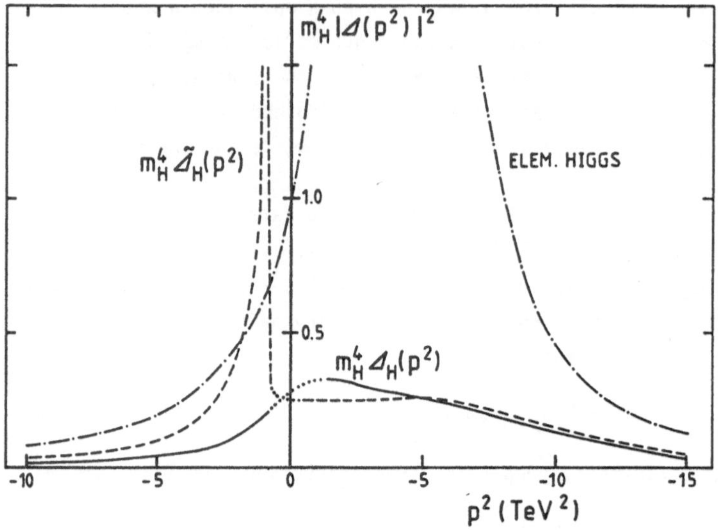

Figure 2.1: Modulus squared of the Higgs boson propagator versus p^2 in the timelike and spacelike regions for $m_H = 2$ TeV. The figure shows the propagator for an elementary scalar (dot-dashed line), the standard-Higgs propagator (dashed line) as well as our modified propagator.

A comparative illustration of the Higgs propagator (case A) is shown in Fig. 2.1, where for $m_H = 2$ TeV the modulus squared of the Higgs-boson propagator is plotted for the three cases of the elementary propagator, $1/(p^2 - m_H^2)$, the standard-model propagator (2.5) and the modified propagator (2.9,10). Introducing the finite width regularizes the elementary propagator in the timelike region, and upon removal of the pole at $p^2 = p_{SP}^2$, the propagator becomes regular in the whole p^2 domain.[1]

It is clear that the regularization of the standard Higgs propagator for $m_H^2 \gg M_W^2$ corresponds to a modification of the original Higgs interaction of the $SU(2)_L \times U(1)_Y$ theory. Further investigations of the full implications of our procedure will be necessary.

3. Heavy Higgs Interactions and Electroweak Precision Experiments at the Z^0 Peak

In principle, the most direct way of exploring empirically the possible existence as well as the properties of a Higgs scalar of mass $m_H > 2M_W$

[1] As our expression for the Higgs propagator is not directly applicable in the threshold region of $p^2 \hat{=} 0$, we have appropriately connected the spacelike and timelike regions by a dotted curve.

consists of measuring the scattering of longitudinal $W^{\pm}$, Z^0 vector bosons on each other /Chanowitz et al., 1985; Kuroda et al., 1988/, where Higgs-boson exchange contributes in the direct and the crossed channels. Obviously, colliders in the multi-TeV-energy range (LHC, SSC or CLIC) are necessary for such experiments.

At present one may attempt to investigate the question of the existence and the properties of the Higgs boson indirectly via virtual effects in high-precision experiments at the presently available energies, in particular, by separately measuring the Z^0 mass and the weak mixing angle, s_W^2, at LEP 100, as well as the $W^{\pm}$ mass at LEP 200 /Proc. ECFA Workshop on LEP 200, 1987/.

The radiative correction, Δr, which enters the Born-term formula for the Z^0 mass via /Sirlin, 1980/

$$M_Z^2 = \frac{\alpha(0)\pi}{\sqrt{2}\,G_F}\;\frac{1}{s_W^2\,c_W^2(1-\Delta r)} \tag{3.1}$$

receives Higgs-boson contributions according to the diagrams of Fig. 1.2. Actually Δr consists of a sum of two parts, a dominating fermionic one, $\Delta r^{(f)}$ /Gounaris et al., 1988/, and a part due to the bosonic vacuum polarization, $\Delta r^{(b)}$, which contains the Higgs-boson contributions of Fig. 2.1, i.e.,

$$\Delta r = \Delta r^{(f)} + \Delta r^{(b)} \quad . \tag{3.2}$$

While $\Delta r^{(f)}$ is strongly dependent on the mass, m_t, of the top quark, $\Delta r^{(b)}$ is essentially (ignoring two-loop effects) independent of m_t.

Suppose the self-energy contribution to the $W^{\pm}$, Z^0 propagator due to the (elementary) Higgs exchange diagrams of Fig. 1.2 is given by $\Sigma_\delta^{(W,Z)}(q^2,y)$, where y stands for the square of the Higgs mass, the modified self-energy incorporating the Higgs propagator (2.10) is obtained via

$$\Sigma^{(W,Z)}(q^2,m_H^2) = \frac{1}{\pi}\int_0^\infty \Sigma_\delta^{(W,Z)}(q^2,y)\;h(y,m_H^2)\;dy \quad . \tag{3.3}$$

The introduction of the modified propagator thus simply amounts to a finite width correction for the Higgs boson, which becomes increasingly more important the larger the mass of the Higgs boson. It is useful to explicitly display the finite-width effect on the vacuum polarization, $\Delta\Sigma^{(W,Z)}(q^2,m_H^2)$, by subtracting the zero-width result $\Sigma_\delta^{(W,Z)}(q^2,y{=}m_H^2)$ from (3.3), i.e.,

$$\Delta\Sigma^{(W,Z)}(q^2,m_H^2) = \frac{1}{\pi}\int_0^\infty \Sigma_\delta^{(W,Z)}(q^2,y)\;\left(h(y,m_H^2) - \pi\,\delta(y-m_H^2)\right)\,dy \quad , \tag{3.4}$$

and to correspondingly split up the bosonic vacuum-polarization correction, $\Delta r^{(b)}$, into a zero-width Higgs-boson part, $\Delta r_\delta^{(b)}$, and a finite-width correction, to be called $\Delta r_{hH}^{(b)}$,

$$\Delta r^{(b)} = \Delta r_0^{(b)} + \Delta r_{hH}^{(b)} \quad , \tag{3.5}$$

where the second term only contributes in the heavy Higgs (hH) limit of $m_H > 1$ TeV. We note that, strictly speaking, due to the problematics of the standard model in the limit of large m_H, only the sum of the two terms in (3.5) is reasonable for $m_H > 1$ TeV.

Substituting the expression for $\Sigma_0^{(W,Z)}$ /Hollik, 1988/ into (3.3) and using the well-known expression connecting $\Delta r^{(b)}$ /Proc. Workshop "Polarization at LEP", 1988; Proc. Workshop on LEP Physics, 1989/ with the vacuum-polarization corrections to the $W^{\pm}$, Z^0 propagators, the (non-standard, heavy Higgs) finite-width correction, $\Delta r_{hH}^{(b)}$, takes the following form

$$\Delta r_{hH}^{(b)}(m_H^2) = \frac{\sqrt{2}\, G_F\, M_W^2}{16\pi^2}\, \frac{11}{3}\, [(S_0-1)\, \ln(\frac{m_H^2}{M^2}) - (S_1-1)\, (\ln(\frac{\Lambda^2}{M^2}) + \frac{5}{6})] \quad . \tag{3.6}$$

where the Higgs-boson spectral-weight function, $h(y,m_H^2)$, is contained in

$$S_0(m_H^2) = \frac{1}{\pi\, \ln(m_H/M^2)} \int_0^\infty h(y,m_H^2)\, \ln(\frac{y}{M^2})\, dy \quad , \tag{3.7}$$

and

$$S_1(m_H^2) = \frac{1}{\pi} \int_0^\infty h(y,m_H^2)\, dy \quad . \tag{3.8}$$

We note that a (mild) logarithmic cut-off dependence has remained in $\Delta r_{hH}^{(b)}(m_H^2)$. Clearly, $\Delta r_{hH}^{(b)}(m_H^2)$ vanishes in the zero-width limit of $h(y,m_H^2) \to \pi\, \delta(y - m_H^2)$.

In Fig. 3.1 we show a plot of $S_0(m_H^2)$ and $S_1(m_H^2)$ as a function of m_H for the two cases of models A and B. The deviation of $S_{0,1}(m_H^2)$ from $S_{0,1} = 1$ measures the change in $\Delta r^{(b)}$ due to the finite-width heavy-Higgs effect compared with the (unrealistic) extension of the zero-width formula $\Delta r_0^{(b)}$ into the region of $m_H \gtrsim 1$ TeV. We note that $S_1 \to 1$ for $m_H \to \infty$ implies that the (logarithmic) dependence of $\Delta r_{hH}^{(b)}$ on Λ disappears in this limit.

A numerical evaluation[2] of $\Delta r_{hH}^{(b)}(m_H^2)$ from (3.6) for the mentioned two cases, called A and B, respectively, yields the results of Table 3.1. According to (3.5) the heavy Higgs-finite-width effect has to be added to the bosonic vacuum-polarization contribution, $\Delta r_0^{(b)}$, which contains the Higgs propagator in zero-width approximation. The values for $\Delta r_0^{(b)}$ are

---

[2] In the numerical evaluation of (3.6) we used $\Lambda = 3m_H$. We have checked that the results are changed insignificantly for different choices of Λ. It has also been checked numerically that our approximation of neglecting the W,Z masses in the expression for the Higgs propagator is perfectly adequate for the results presented in Table 3.1.

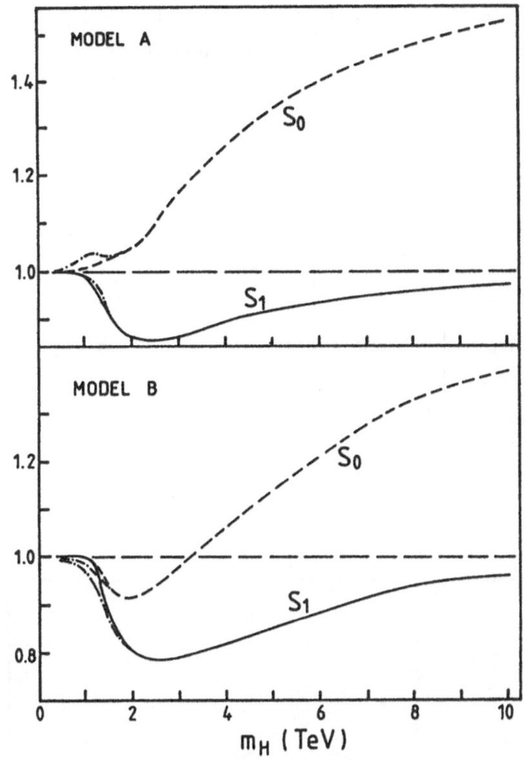

Figure 3.1: The heavy-Higgs large-width correction functions $S_0(m_H^2)$ and $S_1(m_H^2)$ as a function of m_H for Models A and B. The dot-dashed modifications around $m_H \cong 1$ TeV correspond to threshold corrections obtained by appropriately introducing $\xi = 1$ and $M = (M_W + M_Z)/2$.

also quoted in Table 3.1 as well as the sum of the two terms, $\Delta r^{(b)}$. One notes that with increasing mass, m_H, the finite-width effect increases from about 10 % of $\Delta r^{(b)}$ for $m_H = 1$ TeV, to about 75 % for the extreme case of $m_H = 10$ TeV.

The total value of Δr, according to (3.2), is obtained by adding the bosonic contribution, $\Delta r^{(b)}$, of Table 3.1 to the top-mass-dependent fermionic one, $\Delta r^{(f)}$. The values for $\Delta r^{(f)}$, which are identical to the values of Δr evaluated for $m_H = 0.1$ TeV, are given in /Proc. Workshop on LEP Physics, 1989; Gounaris et al., 1988/. The resulting Δr is plotted in Fig. 3.2 as a function of m_t for various values of m_H. Purely for comparison, in Fig. 3.2, we also give the results obtained for Δr by using the elementary-Higgs propagator, i.e., for $\Delta r_{hH}^{(b)} = 0$, even though these values are actually without any significance, as the perturbative (and even the non-

Table 3.1: The results for the finite-width effect, $\Delta r_{hH}^{(b)}$, and the full bosonic vacuum-polarization contribution $\Delta r^{(b)}$ (compare /Proc. Workshop "Polarization at LEP", 1988; Proc. Workshop on LEP Physics, 1989/ for $\Delta r_0^{(b)}$).

m_H (TeV)	0.1	1	2	5	10
$10^3 \cdot \Delta r_{hH}^{(b)}$ A	0	0.6	4.1	10.1	15.0
$10^3 \cdot \Delta r_{hH}^{(b)}$ B	0	0.2	3.1	7.4	11.7
$10^3 \cdot \Delta r_0^{(b)}$	0	10.0	13.0	17.0	20.0
$10^3 \cdot \Delta r^{(b)}$ A	0	10.6	17.1	27.1	35.0
$10^3 \cdot \Delta r^{(b)}$ B	0	10.2	16.1	24.1	31.7

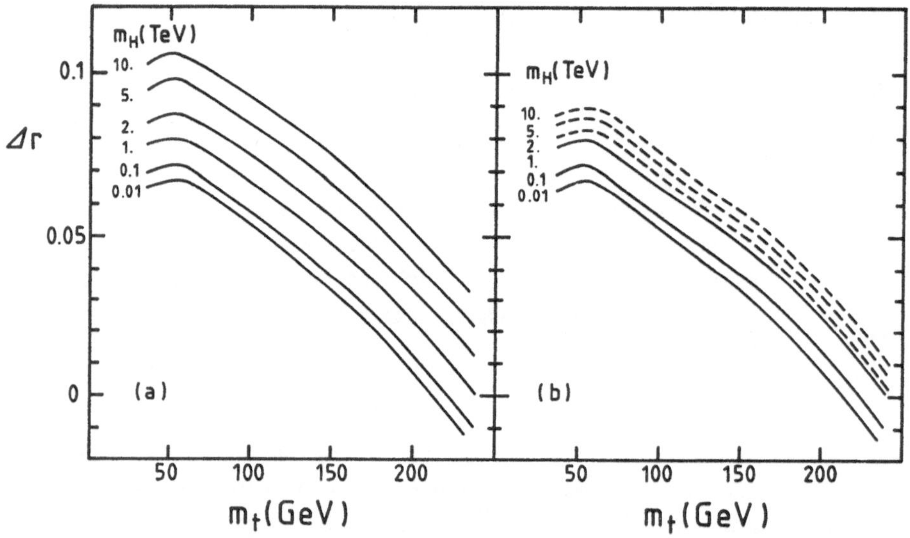

Figure 3.2 a: The radiative correction Δr, evaluated with the modified Higgs propagator taking into account the width of the Higgs boson. Note that for $m_H = 0.1$ TeV the bosonic contribution, $\Delta r^{(b)}$, is negligible /Gounaris et al., 1988a, 1988b/, $\Delta r \equiv \Delta r^{(f)}$.

b: The radiative correction, Δr, if the zero-width formula is evaluated for $m_H > 1$ TeV.

perturbative) unmodified theory becomes unacceptable in the large-m_H limit.

As regards to observability of the bosonic vacuum-polarization contribution, $\Delta r^{(b)}$, we note that the expected accuracy /Proc. Workshop "Polarization at LEP", 1988; Proc. Workshop on LEP Physics, 1989/ of the experiments at LEP 100 ranges from $\delta(\Delta r) = \pm 0.006$ ($\mu^+\mu^-$ pair asymmetry), $\delta(\Delta r) = \pm 0.003$ ($b\bar{b}$ asymmetry) to $\delta(\Delta r) = \pm 0.001$ (polarized beams), while from the $W^{\pm}$-mass determination at LEP 200 one expects $\delta(\Delta r) = \pm 0.005$. A comparison of the expected experimental accuracies with the results in Table 3.1 and Fig. 3.2 shows that one may in fact be able to constrain m_H, or to possibly even isolate a deviation from the case of $m_H = 0.1$ TeV, where Δr is of purely fermionic origin.

At this point, it seems appropriate to add a comment on the possible effect on Δr of additional heavy-vector bosons, which have been conjectured /Casalbuoni et al., 1985, 1987, 1988; Rosenfeld et al, 1988/ to arise in the strongly interacting limit of $m_H \gg 1$ TeV. Indeed, in this limit, one expects also the corrections due to the diagrams on the left-hand side of Fig. 3.3 to become important. As depicted in Fig. 3.3, these diagrams may lead to resonance formation, in particular to vector-boson-resonances, and accordingly, their effect on the vector-boson masses may be described in terms of mixing with the $W^{\pm}$ and Z^0 bosons, the mixing parameters being determined (in principle) by the strong interactions occuring for $m_H \gg M_{W,Z^0}$.

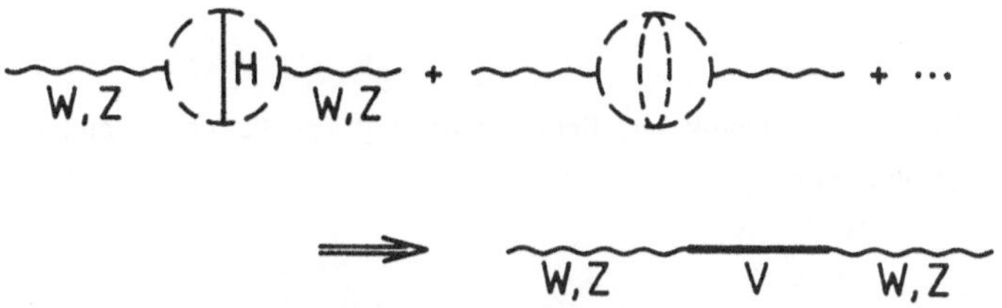

Figure 3.3: Bound-state formation of vector bosons in the limit of m_H becoming large, $m_H \gg 1$ TeV.

A detailed analysis of the effect of such additional vector bosons has to rely on certain model assumptions. In the two models we have investigated, one finds /Gounaris et al., 1989/ that the contribution of heavy-vector states to Δr is either positive, such that it may be neglected, when extracting the most conservative bound on m_H from future data by comparison with the results in Fig. 3.2, or else, that it is negligible small. Of course, additional states of even higher spin may in principle appear, once one enters a strongly interacting regime, but a discussion of such effects is obviously beyond the scope of the present investigation.

4. Conclusions

Starting from the observation that the Higgs-boson propagator in the limit of $m_H \gtrsim 1$ TeV develops a physically unacceptable pole in the space-like region, we propose to introduce a new propagator constructed by conservatively modifying the standard Higgs propagator via imposing decent analyticity requirements. While further theoretical and empirical implications of our approach remain to be investigated, we first of all examined its consequences for the radiative correction, Δr, to be determined via precision electroweak experiments at the Z^0 peak. We find that indeed a heavy-Higgs interaction, corresponding to strong interactions in longitudinal vector-boson scattering, can lead to an appreciable enhancement in Δr. We also noted that this enhancement in Δr may well be much larger than the contribution to Δr of additional narrow resonant states which may arise in the large-m_H limit. Conversely, if an enhancement in Δr will be found (assuming the mass of the top-quark to be known) in nature, it can be explained by our heavy-Higgs extension of the standard formula for Δr.

It is amusing to observe that some of today's burning questions related to the Higgs particle and longitudinal vector-boson scattering physics at the TeV energy scale have much in common with the questions on pion physics at GeV energies studied by ANDRE MARTIN in the sixties /Martin, 1967; Łukaszuk et al., 1967/ in several pioneering papers.

Acknowledgement

One of us (G.G.) would like to thank the Greek Ministry for Research and Technology for partial support.

References

Cabibbo, N., et al. /1979/: Nucl. Phys. B158, 295 (Sects. 1, 2)

Casalbuoni, R., et al. /1985/: Phys. Lett. B155, 95; /1987/: Nucl. Phys. B282, 235; /1988/: Nucl. Phys. B310, 181 (Sects. 1, 3)

Chanowitz, M.S., M.K. Gaillard /1985/: Nucl. Phys. B261, 379 (Sect. 3)

Glashow, S.L. /1961/: Nucl. Phys. 22, 579 (Sect. 1)

Gounaris, G., F.M. Renard, D. Schildknecht /1989/: University of Bielefeld preprint BI-TP 89/41 (Sects. 1, 3)

Gounaris, G.,D. Schildknecht /1988a/: Z. Phys. C40, 477; /1988b/: Z. Phys. C42, 107 (Sect. 3)

Gunion, J.F., et al., /1989/: "The Higgs Hunter's Guide", UCD-89-4 (Sect. 1)

Haber, G.H., G.L. Kane /1985/: Phys. Rep. 117C, 75 (Sect. 1)

Hollik, W.F.L. /1988/: DESY 88-188 (Sect. 3)
Kuroda, M., F.M. Renard, D. Schildknecht /1988/: Z. Phys. C40, 575 (Sect. 3)
Kuti, J. /1988/: Talk at 24th Int. Conf. on High Energy Phys., Munich, p. 814 (Sect. 1)
Lee, B.W., C. Quigg, H.B. Thacker /1979/: Phys. Rev. D16, 1519 (Sect. 1)
Łukaszuk, L., A. Martin /1967/: Nuovo Cim. 52, 122 (Sect. 4)
Lüscher, M., P. Weiß /1988/: Phys. Lett. B212, 472 (Sect. 1)
Maiani, L., et al. /1978/: Nucl. Phys. B136, 115 (Sects. 1, 2)
Marciano, W., G. Valencia, S. Willenbrock /1989/: Phys. Rev. 40, 1725 (Sect. 1)
Martin, A. /1967/: Nuovo Cim. 47A, 265 (Sect. 4)
Proc. ECFA Workshop on LEP 200 /1987/: ed. by A. Bohm, W. Hoogland, CERN 87-08 (Sect. 3)
Proc. Workshop on LEP Physics /1989/: ed. by G. Altarelli, R. Kleiss, C. Verzegnassi, CERN 89-08, Vol. I and II (Sects. 1, 3)
Proc. Workshop "Polarization at LEP" /1988/: ed. by G. Alexander et al., CERN 88-06 (Sects. 1, 3)
Rosenfeld, R., J.L. Rosner /1988/: Phys. Rev. D38, 1530 (Sects. 1, 3)
Salam, A. /1968/: in Elementary particle theory, ed. by N. Svartholm (Almquist and Wiksell, Stockholm) p. 367 (Sect. 1)
Sirlin, A. /1980/: Phys. Rev. D22, 971 (Sect. 3)
t' Hooft, G. /1971/: Nucl. Phys. B33, 173 (Sect. 1)
t' Hooft, G. /1980/: Recent Developments in Gauge Theories, ed. by G. t' Hooft et al. (Plenum Press, New York) (Sect. 1)
Veltman, M. /1977/: Acta Phys. Pol. B8, 475 (Sect. 1)
Veltman, M. /1989/: University of Leiden preprint (May 1989) (Sect. 1)
Weinberg, S. /1967/: Phys. Rev. Lett. 19, 1264 (Sect. 1)

Geometric Approaches to Particle Physics

F. Scheck

Institut für Physik, Johannes-Gutenberg-Universität,
Staudinger Weg 7, Postfach 39 80,
D-6500 Mainz, Fed. Rep. of Germany

Abstract:

Geometric approaches to particle physics have opened up new perspectives and unifying insights. After a few historical remarks I discuss the essence of the concept of G-theory: a primordial symmetry acting on a manifold and on the fields defined on it. This is then illustrated by the finite-dimensional case of Kaluza-Klein theories and by the infinite-dimensional case of chiral anomalies in Yang-Mills theories. In the latter case, a new and unifying description of topological and global anomalies is obtained.

Introduction

In the early 1960's (I was still a student at that time), theoretical particle physics was divided very clearly into phenomenology on one side, and mathematical physics on the other. The former, stimulated by the breath - taking experimental discoveries of the time, was living plainly its "folies de jeunesse", adopting simple, sometimes rudimentary, but naively optimistic approaches. The latter was struggling with topics in general S-matrix theory and axiomatic field theory which often seemed to have little bearing upon the problem of how to explain the "particle zoo". Then and now André Martin belonged to the community of mathematical physicists. But, unlike many of the purists in that community, his motivation in the choice of research topics was always a physical one. His concern was not to prove the mathematical consistency of known empirical procedures but rather to find out to which extent the "rigorous" methods could be given a useful purpose in understanding particle physics. Therefore, André may have been pleased, over the last decade or so, by the cross-fertilization of modern mathematics and elementary particle physics that started with the development of non-Abelian gauge theories. It is in this spirit, and through this short review on geometric approaches to particle physics, that I wish André many more years of fruitful research and fun with physics.

Geometric ideas in describing physics, in some way or other, always arose in relation with the idea of unification. The notions of space-time manifold and of local gauge invariance - the building blocks of modern gauge theories - may be quoted as the first examples. Indeed, the former was fully realized only after the covariance of Maxwell's equations was understood by Minkowski, Einstein, Lorentz and others. Geometrically speaking, it really meant the unification of time and space in a nontrivial way. This is so familiar today that it seems strange to us that it took years before this was fully accepted. Judging from Max Born's recollections of his graduate studies in Göttingen, even as great a mathematician as Felix Klein, by 1909, had not fully realized the unified structure of space-time that follows from special relativity. The second notion, local gauge invariance, which followed from the U(1)-invariance of electrodynamics and which was discovered and clarified by H. Weyl and O. Klein, was again a geometric concept. It turned out to be very fruitful only several decades after its discovery.

Of course, general relativity is an intrinsically, highly nontrivial, geometric theory. It has a unifying character, too, because the metric tensor carries the dynamics and, at the same time, determines the causal structure of space-time. As is well-known, from about 1920 until the late 1960's, there were two lines of attack which developed in parallel but had little influence on each other: the "geometric" approach starting from general relativity trying to give a geometrical interpretation to fundamental interactions beyond gravity, on one side, and the "quantum" approach from quantum mechanics to constructive quantum field theory on the other. On the geometrical side there was the very intriguing work of Th. Kaluza in 1921[1)] that was clarified and further developed by O. Klein in 1926[2)]. It concerned general relativity in 1 time and 4 space dimensions and turned out to be, after projection and compactification to (1+3) dimensions, a unified theory of Einstein's general relativity and electrodynamics, the unification being a genuine geometrical one. The "geometrisation" of all interactions for many years was the main theme of Einstein's scientific work, as well as of others, but it did not reach any conclusive stage.[3] It was only in the mid-sixties, under the impetus of the discoveries in particle physics, that the subject of geometrical theories was taken up again. After it had become clear that symmetries of all types, spectrum symmetries and dynamic symmetries, hidden, broken or unbroken, played an essential role in the physics of elementary particles, geometric theories including fermions (matter fields) and scalars were studied by J.M. Souriau,[4] J. Rayski,[5] A. Trautmann,[6] and many others lateron. From the early 1970's on the geometrical framework of Kaluza-Klein theories had an impressive renaissance (which may be traced back from refs.7) and 8), for instance).

On the "quantum line", geometric aspects made their appearance, at first somewhat implicitly, in the non-Abelian gauge theories, although, for some time at least, one could still live and work with the local infinitesimal (Lie algebra) and algebraic approach that we all had learnt earlier and were used to. The relevance and importance of global geometric aspects became evident, at the latest, when it was realized that anomalies - although they were discovered within perturbation theory - are nonperturbative phenomena which reflect general topological properties of the gauge group and the principal fibre bundle on which the gauge theory is constructed.

From about 1973 on the two lines of approach converged rapidly. On the one hand, the geometric understanding of local gauge theories was fully developed by C.N. Yang, Y.M. Cho and many others. On the other hand, theories of physics in higher dimensions, Kaluza-Klein theories, strings and superstrings, all aimed at unification of the fundamental interactions and made heavy use of differential geometry and topology - both in the formulation of the theory and in developing the concepts of geometric compactification and dimensional reduction to ordinary four-dimensional space-time.

Today, even though many issues of these frameworks are still open, geometric and topological methods have become essential in particle physics. They are useful not only in clarifying the subtleties of chiral anomalies, or the relationship between gravity and the other gauge theories, but will be instrumental, I believe, for a better understanding of spontaneous symmetry breaking and for the tedious struggle for real unification at a fundamental level.

2. Symmetry Action and G-Theory

Symmetries of particle physics usually act in a more subtle way than we thought or learnt in the sixties. A typical situation is this: The theory is invariant under a primordial symmetry which is given by a compact Lie group G. I call it primordial because the effective, observable physics usually has less symmetry than the action from which one starts. G acts on the manifold U on which the theory is formulated, as well as on the geometric objects which live on this manifold and which are the parents of the physical fields of the effective physics. In general, the action of G on U is not free and, therefore, U will be decomposed into strata each of which is characterized by an isotropy group H or I_k, respectively

$$U = U_{(H)} \cup U_{(I_1)} \cup \ldots \cup U_{(I_n)} \tag{1}$$

Here, $U_{(H)}$, $U_{(I_k)}$ is the union of all G-orbits of type (H), or (I_k), respectively, i.e. orbits with stability groups conjugate to H or I_k. The pattern of this stratification of the initial manifold U by G is relevant for the interpretation in terms of physics because it determines the effective dynamics. This will be illustrated by some examples below.

Let I be a fixed isotropy group (H or I_k in eq.(1)). $U_{(I)}$ is a smooth submanifold of U. It also is a fibre bundle over the manifold $M = U_{(I)}/G$, with the fibre G/I, called orbit bundle or stratum. (It is associated to the principle bundle over M with structure group N(I)/I, with N(I) the normalizer of I in G.) There exists a <u>maximal</u> orbit type characterized by the conjugation class (H) whose orbits are diffeomorphic to G/H. The corresponding orbit bundle $U_{(H)}$ is an open and dense submanifold of U; $(U_{(H)}/G$ is connected if $U_{(H)}$ is). Thus $U_{(H)}$ has the same dimension as U. The remaining strata have dimensions smaller than that and are called singular and exceptional. As an example, take $G = U(1)$ acting on $U = \mathbb{C}^2 = \{(u_1,u_2) | u_1,u_2 \in \mathbb{C}\}$ as follows:

For $s \in G$

$$U \times G \longrightarrow U : (u_1,u_2) \cdot s \longmapsto (u_1 s^p,\ u_2 s^q)$$

where p and q are relative prime numbers. Obviously, the origin $(u_1=0,\ u_2=0)$ is a fixed point under G, so its isotropy group is U(1). Consider now the submanifolds $A_p := \mathbb{C}^* \times \{(0)\}$ and $A_q := \{(0)\} \times \mathbb{C}^*$ of U where $\mathbb{C}^*$ is $\mathbb{C}$ from which the origin is subtracted, $\mathbb{C}^* = \mathbb{C} - \{(0)\}$. The isotropy groups of A_p and A_q are $\mathbb{Z}_p$ and $\mathbb{Z}_q$, respectively, because

$$(u_1,0) \cdot s = (u_1 s^p,0) = (u_1,0) \text{ iff } s^p = 1 ,$$

that is, if and only if $s \in \mathbb{Z}_p$, (likewise for A_q). Thus, the U(1) action on $U = \mathbb{C}^2$ stratifies U into

$$U = U_{(\mathbb{1})} \cup U_{(\mathbb{Z}_p)} \cup U_{(\mathbb{Z}_q)} \cup U_{(G)} .$$

$U_{(\mathbb{1})}$ is the principal orbit bundle. It is $\mathbb{C}^2$ minus the origin and minus A_p, A_q and has the structure

$$U_{(\mathbb{1})} = \mathbb{R}^+ \times \mathbb{R}^+ \times S^1 \times S^1 .$$

It has the same dimension as U and its base space is $\mathbb{C}^* \times \mathbb{C}^*/U(1)$. A_p and A_q are exceptional orbit bundles. Their orbits have the same dimension as the orbits of the principal orbit bundle but their base space has a lower dimension. Finally, $U_{(G)} = \{(0,0)\}$ is a singular orbit bundle.

On the manifold, "combed" by the primordial symmetry G as sketched above, physics is formulated in terms of fields Ψ, or cross sections with values in a vector space F, $\Psi : U \to F$. Very much like U, F is a G-space, i.e. the action of G on F is well defined, and the vector bundle $\pi : F \to U$ is a G-vector bundle which means that the projection π is equivariant, $\pi(f\cdot g) = \pi(f)\cdot g$ with $f \in F$, $g \in G$. Usually, only those sections Ψ_Δ are physically relevant which obey a certain G-invariant property Δ. This property, e.g., can be the requirement that Ψ_Δ obey a G-invariant equation of the motion. For example, take $U = \mathbb{R}^{(1,3)}$ to be Minkowski space-time, $F = \mathbb{R}^{(1,3)} \times \mathbb{C}^4$, and Ψ_Δ to be those spinor fields which are solutions of the Dirac equation. Two typical examples will be discussed below, one where U is finite dimensional, and one where it is infinite dimensional. As the construction described above is general and typical for many physical situations, we have proposed to call it a G-theory:[9] A G-theory is a triple

$$(U,\ \Gamma_\Delta F,\ G) \tag{2}$$

of the G-manifold U, the primordial symmetry G, and cross sections $\Psi_\Delta \in \Gamma_\Delta F$ which are charcterized by a G-invariant property Δ.

In fact, one can go one step further. As $U_{(H)}$, the principal orbital bundle is open and dense in U, one can restrict the triple (2) to

$$(U_{(H)},\ \Gamma_\Delta F|_{U(H)},\ G) \tag{3}$$

where $\Gamma_\Delta F|_{U(H)}$ are the corresponding restrictions of the cross sections (with some additional smoothness properties), without loss of information.[9] As the "theory" (3) contains only one type of orbits, it is an obvious candidate for geometric compactification: The physics is G-invariant, every orbit of $U_{(H)}$ can be identified with its base point on the base manifold $U_{(H)}/G$ which is the physically accessible space. All that remains to be done is the reduction of the "parent" fields Ψ_Δ to the effective fields on $U_{(H)}/G$.

3. Kaluza-Klein Theories

The idea of Kaluza-Klein theories is still as intriguing as it was sixty years ago. Its strength and its charm is the unification of both symmetries and interactions in the framework of "simple" physics formulated in a higher-dimensional universe. The distinction between internal and space-time symmetries appears only in the "effective", dynamically complicated physics in four space-time dimensions and stems from the compactification of (n-4) extra dimensions. The complexity and richness of interactions in four dimensions (Einstein gravity, Yang-Mills fields, scalars with nontrivial interactions) is a consequence of dimensional reduction.

These theories are vulnerable, though, and their weaknesses, in a way, are part of their strength: Once the primordial symmetry G, the big universe U, and some matter field(s) on U are given, everything in the effective theory at the level of the four dimensional space-time is fixed and very little room is left in interpreting the effective physics. The identification of the known interactions in M^4 (gravity, electroweak interaction, etc.) and their empirical coupling constants implies compactification of the extra dimensions over very small distances. This means, in turn, that the effective matter fields in M^4 (the "daughters" of the initially massless "parents" on U) aquire large masses, of the order of the Planck scale $\sqrt{\hbar c/G_{Newton}} \simeq 10^{19}\mathrm{GeV}/c^2$, too large for any obvious interpretation in particle physics. All models which yield the structure group of the minimal model in M^4, predict a variety of scalars with fixed and nontrivial couplings. So, in a way, they have too much predictive power.[7),8),10)]

A nice albeit somewhat academic example for a Kaluza-Klein theory in the geometric framework of G-theory is provided by the monopole solutions of five dimensional Einstein equations.[11)-13)] The primary, "big" universe has the structure $U = \mathbb{R}_t \times \mathbb{R}^4 = \mathbb{R}^5$. A static magnetic monopole is described by a metric of the form

$$g^{(u)} = \begin{pmatrix} 1 & 0 \\ 0 & \gamma \end{pmatrix} , \qquad (4)$$

where γ is the self-dual instanton metric.[14)] Among the isometries of the metric it is the symmetry G = U(1) that matters for the compactification. Indeed, G stratifies U into the principal orbit bundle

$$U_{(H)} = \mathbb{R}_t \times (\mathbb{R}^4 - \{0\}) \tag{5}$$

and the singular orbit bundle $\mathbb{R}_t \times \{0\}$. The latter, when projected to four dimensions, is nothing but the world line of the monopole, i.e. the source of the monopole field. The former, $U_{(H)}$, is the natural candidate for Kaluza-Klein space that compactifies spontaneously to[15)]

$$M^4 = U_{(H)}/G = U_{(H)}/U(1) \ . \tag{6}$$

$U_{(H)}$, when probed <u>locally</u>, has the structure $M^4 \times U(1)$. When looked at <u>globally</u>, however, it appears to be twisted by the presence of the monopole and does not have the trivial structure of M^4 multiplied by the internal space $U(1) \cong S^1$. The Kaluza-Klein space $U_{(H)}$, eq.(5), can be expressed in terms of polar coordinates $\{r \in \mathbb{R}^+;\ (\Theta,\Phi,\Psi) \in S^3\}$, so that

$$U_{(H)} = \mathbb{R}_t \times \mathbb{R}^+ \times S^3 \ . \tag{7}$$

With the U(1) action, the geometrically essential part of this is the Hopf bundle

$$\begin{array}{ccc} S^1 & \longrightarrow & S^3 \\ & & \downarrow \\ & & S^2 \end{array} \ .$$

If (for fixed time t and radial variable r) the S^3 is parametrized by $|z_1|^2 + |z_2|^2 = 1$, $z_1, z_2 \in \mathbb{C}$, and if it is described locally by $S^2 \times S^1$, then the charts on the northern hemisphere of the S^2 (coordinates Θ, Φ, Ψ_+) and on its southern hemisphere (Θ, Φ, Ψ_-) are related by the transition map at $\Theta = \pi/2$ (for a monopole of magnetic charge 1)

$$e^{i\Psi_-} = e^{i\Phi}\, e^{i\Psi_+} \ . \tag{8}$$

On the two hemispheres the metric, as induced by the monopole, has the form

$$\begin{aligned} ds^2 &= -dt^2 + \lambda^2\, V(r)\, (d\Psi_\pm \pm \tfrac{1}{2}\,(1 \mp \cos\Theta)\, d\Phi)^2 \\ &+ \frac{1}{V(r)}\, (dr^2 + r^2\, (d\Theta^2 + \sin^2\Theta\, d\Phi^2)) \ . \end{aligned} \tag{9}$$

Here $\lambda = 4\sqrt{\pi G_N}/e$, $V(r) = 1/(1 + \frac{\lambda}{2r})$, (r,Θ,Φ) are polar coordinates in physical space, while Ψ is chosen to be 2π-periodic and corresponds to the

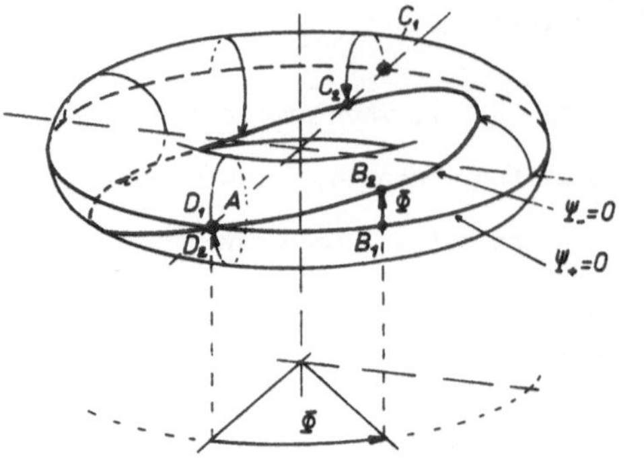

Fig. 1. The circle $AB_1C_1D_1$ represents the restriction of the local 1-cross-section ($\Psi^+ = 0$) corresponding to the chart on the northern hemisphere, while the curve $AB_2C_2D_2$ (with $D_2 \equiv D_1$) corresponds to the chart on the southern hemisphere. The second curve winds once around the torus, illustrating the twist.

Killing vector ∂_Ψ related to the G-invariance of the metric. The twist contained in (8) is illustrated in Fig.1 which shows how the lines $\Psi_+ = 0$ and $\Psi_- = 0$ must be joined at the equator ($\Theta = \pi/2$) of the S^2.

The dimensional reduction to M^4 yields the tensor field $g^{\mu\nu}$ of Einstein gravity, the vector potential A_μ, and the scalar field $V^{1/2}$. A_μ is found to be, on the northern and southern hemispheres, respectively,

$$A_\mu = \{A_0=0,\ A_r=0,\ A_\Theta=0,\ A_\Phi^{(\pm)} = \pm\frac{1}{2}(1 \mp \cos\Theta)\} \qquad (10)$$

and corresponds indeed to the magnetic field $\vec{B} = \vec{r}/2r^3$.

The model is nice because it is completely transparent. It illustrates the way the primordial symmetry acts on the primary, multidimensional universe U by preparing the Kaluza-Klein space $U_{(H)}$ (the principal orbit bundle) which compactifies quasi automatically. It is natural that one must cut out of U the submanifold $\mathbb{R}_t \times \{0\}$ before compactification: this is the world line of the monopole. Like any other pointlike source this is a singularity that must be treated separately. Here, in fact, it can be brought back to M^4, after compactification, so that the singular orbit bundle has a physical interpretation. (In the more general case (1) and for higher dimensional models the interpretation of the other strata is not understood, however.) Finally, the geometric approach to the monopole shows in a transparent manner the importance of global topological aspects and provides a new, geometric understanding of Dirac's monopole, eq.(10).

4. Chiral Anomalies and Geometry of Yang-Mills Theories

It is well-known that gauge theories are intrinsically geometric theories. Geometric understanding of a theory always includes its global topological aspects. This is certainly true for Yang-Mills theories even though any practical treatment of their quantized version relies on (field theoretic) locality and covariant perturbation theory, i.e. on an algebraic and local "chartwise" approach. This would probably be the whole story, from a practical point of view, if there was not the tantalizing phenomenon of anomalies. Anomalies represent a specific way of symmetry breaking, in the transition from a classical theory to its quantum analogue. As such, they concern also topological aspects of the theory and, in fact, are intimately connected with the quantization procedure proper. The interesting observation is that the analysis of chiral anomalies is a rather natural, albeit technically complex, application of the concept of a G-theory which clarifies a different facet of anomalies. This is what I wish to sketch in this section.

The gauge theory is formulated on a principle fibre bundle P(M,G) over the space-time manifold M, G being the structure group. The choice of M and, therefore, of P reflects the physical boundary conditions imposed on the gauge and matter fields. Let $\mathcal{A}$ be the space of connections on P and let $\mathcal{G}$ denote the gauge group, i.e. the group of vertical automorphisms on P, $Aut_v P$. Both $\mathcal{A}$ and $\mathcal{G}$ are infinite dimensional. $\mathcal{A}$ is an affine space. The action of the gauge group on this space, in general, is highly nontrivial, and $\mathcal{M} := \mathcal{A}/\mathcal{G}$, the space of physical, gauge-inequivalent connections is a rather complicated object. In general, $\mathcal{M}$ is not a manifold. Therefore, the physics being formulated by means of the action functional on $\mathcal{A}$ (integrated over the chiral fermions), the division by the gauge group $\mathcal{G}$ may lead to problems with the effective functional. The dividing out of the gauge degrees of freedom must be done more carefully.[16)]

For this it is useful to study first the stratification of the space of connections by the action of the gauge group, i.e., in analogy to (1),

$$\mathcal{A} = \mathcal{A}^{(J_o)} \cup \mathcal{A}^{(J_1)} \cup \ldots \cup \mathcal{A}^{(J_k)} \tag{11}$$

Here, the connections $A \in \mathcal{A}^{(J_i)}$ which belong to the stratum of type (J_i) are characterized by their stability group $\mathcal{G}_A$ being conjugate to J_i,

$$\mathcal{A}^{(J_i)} = \{A \in \mathcal{A} \mid \mathcal{G}_A = \Psi J_i \Psi^{-1},\ \Psi \in \mathcal{G}\}, \tag{12}$$

J_i being isomorphic to a subgroup of the structure group G. The number of strata is countable. The main stratum (J_o) has the "symmetry" J_o isomorphic to the center C of G. It is dense in $\mathcal{A}$. It should be noted that the stratification (11) is natural and unique, once the theory is defined by giving P(M,G). Except for the main stratum, however, the consequences of this "combing" of $\mathcal{A}$ for the physics of gauge theories is not well understood as yet.[17] Formally, defining

$$\mathcal{M}_i := \mathcal{A}^{(J_i)}/\mathcal{G}\ , \tag{13}$$

we recover the orbit bundle decomposition that we described in sec.2 and the compactification pattern of Kaluza-Klein theories. The $\mathcal{M}_i$ are pieces of the space of physical, gauge-inequivalent connections but, as we said above, they do not join smoothly to a well-behaved manifold $\mathcal{M}$.

The physics of gauge fields and chiral fermions is defined, at the classical level, by the action $S(A,\bar{\chi},\chi)$ which, of course, is strictly gauge invariant. At the quantum level the relevant quantity is the generating functional, obtained by integrating over the fermionic degrees of freedom,

$$Z(A) = \int [\mathcal{D}\chi]\,[\mathcal{D}\bar{\chi}]\ e^{-S-\bar{\chi}\partial\!\!\!/_A\chi}\ . \tag{14}$$

There are then two possibilities. With $\Psi \in \mathcal{G}$, either Z is strictly invariant, $Z(\Psi\cdot A) = Z(A)$, or it is equivariant but not strictly invariant, $Z(\Psi\cdot A) = \rho(A,\Psi)^{-1}\,Z(A)$, where ρ represents the action on $\mathbb{C}$ which depends on A. In the first case, the division with $\mathcal{G}$ poses no problem. The theory is free of anomalies and can be reduced to $\mathcal{M}$, the space of gauge-inequivalent connections. In the second case, however, anomalies are present and the reduction cannot be performed. In this sense, chiral anomalies may be understood in a geometrical language as an obstruction to the reduction procedure, the obstruction being due to quantization.

Things become simpler if instead of the full gauge group $\mathcal{G}$, (for the non-Abelian case), one divides by the pointed gauge group $\mathcal{G}^*$. This is a subgroup of $\mathcal{G}$ and is defined to be the stability group of an arbitrary but fixed point of the principal fibre bundle P(M,G),

$$\mathcal{G}^* = \mathcal{G}_{p_0} = \{\Psi \in \mathcal{G} \mid \Psi(p_0) = p_0\} ,$$

(it is the identity in the fibre over p_0). The action of $\mathcal{G}^*$ on $\mathcal{A}$ is free[18] and, therefore,

$$\begin{array}{ccc} \mathcal{G}^* & \longrightarrow & \mathcal{A} \\ & & \downarrow \\ & & \mathcal{M}^* = \mathcal{A}/\mathcal{G}^* \end{array}$$

is a principal fibre bundle, $\mathcal{M}^*$ now being a decent manifold.

The functional Z(A) is a trivial section

$$Z : \mathcal{A} \longrightarrow \mathrm{Det} := \mathcal{A} \times \mathbb{C} \tag{15}$$

in the determinant bundle. Dividing with $\mathcal{G}^*$ gives the reduced section

$$Z^* : \mathcal{A}/\mathcal{G}^* \longrightarrow (\mathcal{A} \times \mathbb{C})/\mathcal{G}^* =: \mathrm{Det}^* \tag{16}$$

Now, if Z is strictly invariant, then the action of $\mathcal{G}^*$ on $\mathbb{C}$ is trivial and (16) reduces to

$$Z^* : \mathcal{M}^* \longrightarrow \mathcal{M}^* \times \mathbb{C}, \quad \mathcal{M}^* := \mathcal{A}/\mathcal{G}^* .$$

Det* is trivial, the theory is reducible to $\mathcal{M}^*$. If Z is equivariant but not strictly invariant, then the action of $\mathcal{G}^*$ on C is not trivial. In this case Det* has a twist and integration over $[\mathcal{D}A]$ is not possible. In other words, there is a $\mathcal{G}^*$ -, or topological, anomaly.

Even if no such anomaly is present, this analysis is incomplete insofar as it left out the remainder $\mathcal{G}/\mathcal{G}^*$ of $\mathcal{G}$ which is isomorphic to G, the structure group. This finite-dimensional piece is important (and may hide surprises) because it is the structure group G which defines the physical charges of the particles in the theory. Again, if Z* is strictly invariant, the final division with G poses no problem. If it is only equivariant, however, one obtains*)

$$Z^{**} : \mathcal{M} \longrightarrow (\mathcal{M}^* \times \mathbb{C})/G =: \mathrm{Det}^{**} . \tag{17}$$

Det** is then nontrivial and we have found a G-anomaly. (If $\mathcal{M}$ is not a manifold, restrict the argument to its main stratum $\mathcal{M}_0$.) In this approach, chiral gauge anomalies manifest themselves through the loss of strict invariance for the generating functional. This is more general than manifestation through topological properties, as probed by the index theorems of Atiyah and Singer.[19),18)]

An example of an equivariant anomaly which is not topological is provided by the center anomaly.[17),20)] In the analysis sketched above it appears automatically in the stepwise reduction of the original configuration space $\mathcal{A}$, thus illustrating the unifying procedure of the geometric, G-theory approach.

We restrict the analysis to $\mathcal{A}^{(J_0)} = \mathcal{A}^{(C)}$, the main stratum characterized by the stability group C = center of G. As all connections A of $\mathcal{A}^{(C)}$ are C-invariant by definition, also the image of Z(A) must consist of fixed points of C. This can happen in two ways: either Z(A) is strictly invariant or, if it is only equivariant, its image in C must be the point $\{0\}$. In the latter case there appears an anomaly, the so-called center anomaly which is a global one but not a topological one. The actual computation of Z is done by means of the index theorem. One finds for $c \in C$, and $z \in \mathrm{Det}$

$$cz = \lambda^{\mathrm{ind}\not{\partial}_A} z$$

where λ is a complex number with modulus 1, defined by $\rho(c)\, \chi_n = \lambda\, \chi_n$, χ_n being the spinor fields spanning the kernel of the Weyl-Dirac operator $\not{\partial}_A$[20)]. As an example consider the case of G = SU(2) as the structure group. Here the center is

---

*) At this point it is assumed that there are no center anomalies. If they appear, they must be discussed separately (see below).

$$C(G) = \{\mathbb{1}, -\mathbb{1}\} = \mathbb{Z}_2$$

and $\lambda = +1, -1$. In case $(\mathrm{ind}\,\mathcal{D}_A)$ is even (0,2,4, ...) there is no anomaly. If $(\mathrm{ind}\,\mathcal{D}_A)$ is odd (1,3,5, ...) we get stuck with a global, nonperturbative anomaly. For the example of SU(2) this is the global anomaly of Witten.[21] Here it is identified as an equivariance, or center, anomaly, in a unified, intrinsically geometric approach.

5. Conclusion

Geometric approaches to particle physics reveal new facets which are not only fascinating but are likely to lead to new insights. Somewhat like for modern developments in classical mechanics qualitative and global aspects become important and help to clarify features of the theory which are often hard, if not impossible, to disentangle in the local, algebraic approach that we learnt at school. In relation with gauge theories, the analysis of anomalies has brought forth a deeper understanding of their topological nature but at present it is by far not concluded. Gauge theories are highly nonlinear dynamical theories and they may well give us more surprises in the future concerning some new, unexpected non-perturbative features. If this expectation is justified then, very likely, the way to them will be a geometric one. Yang-Mills theories have had a profound influence on modern differential topology. I am sure that this field of mathematics will pay back to particle physics.

References

1. Th. Kaluza, Sitz.Preuß.Akad.Wiss.Berlin, Math.Phys.Kl.(1921)966
2. O. Klein, Z.Physik 37(1926)895
3. A. Pais: "Subtle is the Lord ..., The Life and Science of Albert Einstein", Clarendon Press, Oxford (1982)
4. J.M. Souriau, Nuovo Cim. 30(1963)565
5. J. Rayski, Acta Phys. Pol. 27(1965)89
6. A. Trautmann, Rep.Math.Phys. 1(1970)29
7. see e.g. the review by J.M. Duff, B.E.W. Nilsson and C.N. Pope, Phys. Reports 130(1986)1

8. R. Coquereaux and A. Jadzcyk; "Riemannian Geometry, Fiber Bundles, Kaluza-Klein Theories and All That"; World Scientific, Lecture Notes in Physics (1988)
9. A. Heil, N.A. Papadopoulos, B. Reifenhäuser and F. Scheck; Nucl.Phys. B293(1987)445
10. A. Heil, N.A. Papadopoulos and F. Scheck; "Reduction mechanism for symmetric fields on an orbit bundle", preprint MZ-TH/88-07, to be published
11. D.J. Gross and M.J. Perry; Nucl.Phys. B226(1983)29
12. R.D. Sorkin; Phys.Rev.Lett. 51(1983)87
13. D. Pollard; J. Phys. A16(1983)565
14. E.T. Newman, L. Tamburino and T. Unti; J.MathPhys. 4(1963)915; A. Taub and C.W. Misner; Zh.Eksp.Teor.Fiz. 55(1968)233; S.W. Hawking; Phys.Lett. A60(1977)81
15. A. Heil, N.A. Papadopoulos, B. Reifenhäuser and F. Scheck; Phys.Lett. B173(1986)149 and Nucl.Phys. B281(1987)426
16. A. Heil, A. Kersch, N.A. Papadopoulos, F. Scheck and H. Vogel; Journ.Geom. and Phys. 6(1989)237; A. Heil, A. Kersch, N.A. Papadopoulos, B. Reifenhäuser and F. Scheck, Ann. of Phys. (N.Y.), in print
17. A. Heil, A. Kersch, N.A. Papadopoulos, B. Reifenhäuser and F. Scheck, "Structure of the space of reducible connections for Yang-Mills theories", subm. to Rev. in Math.Phys.
18. I.M. Singer, Soc. Mathématique de France, Astérisque (1985)323
19. M.F. Atiyah and I.M. Singer, Proc.Nat.Acad.Sci. (USA) 81(1984)2597
20. A. Kersch: "Anomalien im Rahmen des Indextheorems", thesis, Mainz Jan.1990
21. E. Witten, Phys.Lett. B117(1982)324

Index of Contributors

Subject Index

Springer Tracts in Modern Physics

90* **Jets of Hadrons**
By W. Hofmann 1981. 165 figs. VIII, 215 pages

91 **Structural Studies of Surfaces**
By K. Heinz, K. Müller, T. Engel, and K. H. Rieder
1982. 120 figs. VII, 180 pages

92 **Single-Particle Rotations in Molecular Crystals**
By W. Press 1981. 53 figs. IX, 129 pages

93 **Coherent Inelastic Neutron Scattering in Lattice Dynamics**
By B. Dorner 1982. 47 figs. VIII, 96 pages

94 **Exciton Dynamics in Molecular Crystals and Aggregates**
By V. M. Kenkre and P. Reinecker 1982. 37 figs. IX, 226 pages

95 **Projection Operator Techniques in Nonequilibrium Statistical Mechanics**
By H. Grabert 1982. 4 figs. X, 164 pages

96 **Hyperfine Structure in 4d- and 5d-Shell Atoms**
By S. Büttgenbach 1982. 14 figs. VIII, 97 pages

97 **Elements of Flow and Diffusion Processes in Separation Nozzles**
By W. Ehrfeld 1983. 73 figs. VIII, 140 pages

98 **Narrow-Gap Semiconductors**
With contributions by R. Dornhaus, G. Nimtz, and B. Schlicht
1983. 2nd printing 1985. 244 figs, X, 309 pages

99 **Dynamical Properties of IV–VI Compounds**
With contributions by H. Bilz, A. Bussmann-Holderer, W. Jantsch, and P. Vogl
1983. 47 figs. VIII, 101 pages

100* **Quarks and Nuclear Forces**
Edited by D. C. Fries and B. Zeitnitz 1982. 69 figs. XI, 223 pages

101 **Neutron Scattering and Muon Spin Rotation**
With contributions by R. E. Lechner, D. Richter, and C. Riekel
1983. 118 figs. IX, 229 pages

102 **Theory of Jets in Electron-Positron Annihilation**
By G. Kramer 1984. 86 figs. VII, 140 pages

103 **Rare Gas Solids**
With contributions by H. Coufal, E. Lüscher, H. Micklitz, and R. E. Norberg
1984. 40 figs. IX, 99 pages

104 **Surfaces Enhanced Raman Vibrational Studies at Solid/Gas Interfaces**
By I. Pockrand 1984. 60 figs. IX, 164 pages

105 **Two-Photon Physics at e^+e^- Storage Rings**
By H. Kolanoski 1984. 123 figs. IX, 187 pages

106 **Polarized Electrons at Surfaces**
By J. Kirschner 1985. 57 figs. VIII, 158 pages

* denotes a volume which contains a Classified Index starting from Volume 36